SpringerBriefs in Applied Sciences and Technology

PoliMI SpringerBriefs

More information about this subseries at http://www.springer.com/series/11159
http://www.polimi.it

Grazia Concilio · Paola Pucci · Lieven Raes ·
Geert Mareels
Editors

The Data Shake

Opportunities and Obstacles for Urban Policy
Making

POLITECNICO
MILANO 1863

Editors
Grazia Concilio
DASTU
Politecnico di Milano
Milan, Italy

Paola Pucci
DASTU
Politecnico di Milano
Milan, Italy

Lieven Raes
Digitaal Vlaanderen
Brussels, Belgium

Geert Mareels
Digitaal Vlaanderen
Brussels, Belgium

ISSN 2191-530X   ISSN 2191-5318   (electronic)
SpringerBriefs in Applied Sciences and Technology
ISSN 2282-2577   ISSN 2282-2585   (electronic)
PoliMI SpringerBriefs
ISBN 978-3-030-63692-0   ISBN 978-3-030-63693-7   (eBook)
https://doi.org/10.1007/978-3-030-63693-7

This Springer imprint is published by the registered company Springer Nature Switzerland AG
The registered company address is: Gewerbestrasse 11, 6330 Cham, Switzerland

Preface

The book investigates the operative and organizational implications related to the use of the growing amount of available data on policy making processes, highlighting the experimental dimension of policy making that, thanks to data, proves to be more and more exploitable toward more effective and sustainable decisions. According to well-established literature and some empirical evidence, data has shown great potential in supporting the reading and the estimation of fine-grained variations in human practices in cities over time and spaces, depending on the specific sources from which it can be obtained. These features also affect data usability, highlighting how its potential relevance may vary in each different stage of the policy making process. The larger and larger availability of data offers support for (1) strategic activities, by aggregating information on time series that inform and validate public actions; (2) tactical decisions, conceived as the evidence-informed actions that are needed to implement medium-term strategic decisions; (3) operational decisions, giving support to day-by-day decision-making activities in a short-term perspective. The larger the availability of data, the bigger the chance it offers to implement experimental activities feeding the policy design and implementation as well as enabling collective learning processes. Within this changing but promising landscape, the book is structured into two parts.

The first part introduces the key questions highlighted by the PoliVisu project and still representing operational and strategic challenges in the exploitation of data potentials in urban policy making. It includes four chapters.

The chapter titled "The Data Shake: An Opportunity for Experiment-Driven Policy Making" by Grazia Concilio and Paola Pucci starts from considering that the wider availability of data and the growing technological advancements in data collection, management, and analysis question the very basis of the policy making process toward new interpretative models. By dealing with the operative implications in the use of a growing amount of available data in policy making processes, this chapter starts discussing the chance offered by data in the design, implementation, and evaluation of a planning policy, with a critical review of the evidence-based policy making approaches; then introduces the relevance of data in the policy design experiments and the conditions for its uses.

The second chapter by Nils Walravens, Pieter Ballon, Mathias Van Compernolle, Koen Borghys, titled "Data Ownership and Open Data: The Potential for Data-Driven Policy Making," discusses the issue of data ownership within the broader framework of Smart Cities initiatives. It acknowledges that the growing power acquired by the data market and the great relevance assigned to data ownership, rather than to data-exploitation know-how, is affecting the development of a data culture and is slowing down the embedding of data-related expertise inside public administrations. Some key questions are explored in order to assess what are the consequences of these phenomena when imagining the potential for policy making consequent to the growing data quantity and availability; which strategic challenges and decisions do public authorities face in this regard; what are valuable approaches to arm public administrations in this "war on data." By looking at the Smart Flanders program initiated by the Flemish government (Belgium) in 2017, the chapter offers insight to support cities with defining and implementing a common open data policy.

In the chapter "Towards a Public Sector Data Culture: Data as an Individual and Communal Resource in Progressing Democracy," Petter Falk argues that in consequence of the increased use of data, citizens and governments, as both producers and consumers of data, become intertwined in more and more complex ways. For ensuring a democratic usage of citizens data, Falk highlights the relevance of a sound data practice and data culture, drawing on insights from the project Democracy Data, that explores the opportunities and obstacles for establishing democratically oriented public sector data cultures.

In the final chapter of the first book part, titled "Innovation in Data Visualisation for Public Policy Making," Paolo Raineri and Francesco Molinari propose a reflection on the potential of data visualization technologies for (informed) public policy making in a growingly complex and fast-changing landscape. Based on the results of an online survey of more than 50 data scientists from all over the world, the authors highlight five application areas that, according to the domain specialists, see the biggest needs for innovation. In particular, the experts argue, that the transformation of the business cases supporting the adoption and implementation of data visualization methods and tools in government is not fully captured by the conventional view of the value of Business Intelligence. Finally, the authors reflect on citizen science, design thinking, and accountability as triggers of civic engagement and participation that can bring a community of "knowledge intermediaries" into the daily discussion on data-supported policy making.

The second part of the book, focusing on the findings of the Horizon 2020 project PoliVisu, explores the ways data (different in sources as well as in the information and knowledge it activates) and data visualizations (necessary for any discussion or reflection on problems, impacts, or evidences) are changing policy making and especially mobility policy (from the real-time monitoring of traffic to strategic mobility plans) and the different roles they may play in the stages of a policy cycle. Policy cycle model is assumed as an ideal typical model to organize and systematize the phases of a policy making process, and is empirically experienced into project pilot cases (mainly Ghent in Belgium, Issy-les-Moulineaux in France, and Pilsen in Czechia), in order to focus on the experiential dimension of policy making in concrete decision-making contexts.

Yannis Charalabidis opens the second part of the book with his contribution titled "Policy-Related Decision Making in a Smart City Context: The PoliVisu Approach" by considering that ICT provides new methods and tools to politicians and their cabinets on an almost daily basis, so allowing them to deal with the growing quest for more efficient governance. The chapter introduces the perspective of the PoliVisu project that constitutes a step forward from the evidence-based decision making, going toward an experimental approach, supported by the large variety of available data sets. Yannis Charalabidis argues that, by utilizing advanced data gathering, processing, and visualization techniques, the PoliVisu platform is one of the most recent integrated examples promoting the experimental dimension of policy making at a municipal and regional level.

The following chapter titled "Turning Data into Actionable Policy Insights" introduces the project and its main activities. This contribution by Jonas Verstraete, Freya Acar, Grazia Concilio, and Paola Pucci highlights the key impacts and opportunities offered to policy making by the greater availability of data. This analysis is carried out by mapping the key data sources and data analytical approaches entering policy making, along through the different steps of a policy cycle model and underlines the crucial work developed by both policy makers and data specialists associated with data selection, evaluation, and use. The entire chapter benefits from short narrations reporting "data at work" examples, supplied by the several project pilots.

In the chapter titled "Data-Related Ecosystems in Policy Making. The Polivisu Contexts," Giovanni Lanza explores the complexity of the ecosystems that develop around data-supported policy making. Such complexity can be traced back to the multiplicity of actors involved, the roles they assume in the different steps of the decision-making process, and the nature of the relationships they establish, takes on new connotations following the rising use of data for public policies. Issues related to data ownership and the ability to collect, manage, and translate data into useful information for policy makers require, following Lanza, the involvement of several actors, generating ecosystems where co-creation strategies are confronted with the limits of action of the public administrations within broader social and decisional networks. Based on this background, this chapter provides, through the analysis of the direct experiences conducted by the pilot cities involved in the PoliVisu project, an overview of the opportunities and challenges related to the impact of data in the evolution of decision-making networks and ecosystems in the data shake era.

The last chapter, "Making Policies with Data: The Legacy of the PoliVisu Project" by Freya Acar, Lieven Raes, Bart Rosseau, Matteo Satta discusses four questions driving the entire project and the interpretation of the key findings: What are the new roles data can play in the policy making process? What is the added value of data for policy making? How can innovative visualizations contribute to improve the use of data in policy making processes? To what extent can an increased adoption of data affect the policy making process? How is the data shake affecting the involvement of non-institutional actors in the policy making process? After discussing these questions, this last chapter explores some bottlenecks and key lessons learnt in the exploitation of data potentials in policy making.

Milan, Italy   Grazia Concilio
Milan, Italy   Paola Pucci
Brussels, Belgium   Lieven Raes
Brussels, Belgium   Geert Mareels

Contents

Editors and Contributors

About the Editors

Grazia Concilio is Associate Professor in Urban Planning and Design at DAStU, Politecnico di Milano. She is an engineer and Ph.D. in "Economic evaluation for Sustainability" from the University of Naples Federico II. She carried out research activities at the RWTH in Aachen, Germany (1995), at IIASA in Laxenburg, Austria (1998) and at the Concordia University of Montreal, Canada, (2002); she is reviewer for several international journals and former member (in charge for the assessment of LL new applications) of ENoLL (European Network of open Living Lab). Team members in several research projects; responsible for a CNR research program (2001) and coordinator of a project funded by the Puglia Regional Operative Programme (2007–2008) and aiming at developing an e-governance platform for the management of Natural Parks. She has been responsible on the behalf POLIMI of the projects Peripheria (FP7), MyNeighbourhood|MyCity (FP7), Open4Citizens (Horizon 2020); she is currently responsible for the Polimi team for the projects Designscapes (Horizon 2020 www.designscapes.eu), Polivisu (Horizon 2020 www.polivisu.eu) together with Paola Pucci, and MESOC (Horizon 2020 www.mesoc-project.eu). She is coordinating the EASYRIGHTS project (Horizon 2020 www.easyrigths.eu). She is the author of several national and international publications.

Paola Pucci is full Professor in Urban planning and former Research Director of the Ph.D. course in Urban Planning Design and Policy (UPDP) at the Politecnico di Milano. From 2010 to 2011, she taught at the Institut d'Urbanisme in Grenoble Université Pierre Méndes France at Bachelor, Master, and Ph.D. levels and currently visiting professor at European universities. She has taken part, also with roles of team coordinator, in national and international research projects funded on the basis of a competitive call, dealing with the following research topics: Mobility policy and transport planning, mobile phone data and territorial transformations and including EU ERA-NET Project "EX-TRA – EXperimenting with city streets to TRAnsform urban mobility" (ongoing); H2020—SC6-CO-CREATION-2016-2017

"Policy Development based on Advanced Geospatial Data Analytics and Visualisation," EU Espon Project, PUCA (Plan, urbanisme, architecture) and PREDIT projects financed by the Ministère de l'Ecologie, du Développement et de l'Aménagement durable (France). She has supervised and refereed different graduate, postgraduate, and Ph.D. theses at Politecnico di Milano, Université Paris Est Val de Marne, Ecole Superieure d'Architecture de Marseille, Université de Tours. She has been Member of the evaluation panel for The Netherlands Organisation for Scientific Research (NWO, 2017), and Member of the NEFD Policy Demonstrators commissioning panel for the ESRC_Economic and Social Research Council. Shaping Society (Uk), on the topic "New and Emerging Forms of Data—Policy Demonstrator Projects" (2017).

Lieven Raes holds master degrees in Administrative Management and land-use planning. Lieven is a public servant at Information Flanders (Flemish government) and is currently the consortium coordinator of two EU H2020 research and innovation projects regarding the relationship between data, policy making in a smart city context (PoliVisu and Duet).

Before Lieven coordinated the CIP Open Transport Net project (OTN) and participated in several other EU projects (CIP, CEF, FP7, FP5, and FP4) and also in several Flemish ICT, E-Government and smart-city related projects.

Geert Mareels is born in 1962, holds master degrees Administrative Management and Political Science. He leads the eGovernment service of the Flemish Region. He is now project leader for the Digital Archives of Flanders and Coordinator for the implementation of the EU Single Digital Gateway. He was chairman of the Flemish Privacy Commission and was project coordinator of three EU projects.

Contributors

Freya Acar Dienst Data En Informatie, Bedrijfsvoering, Stad Gent, Belgium

Pieter Ballon Imec-SMIT, Vrije Universiteit Brussel, Brussels, Belgium

Koen Borghys Imec-SMIT, Vrije Universiteit Brussel, Brussels, Belgium

Yannis Charalabidis University of the Aegean, Samos, Greece

Mathias Van Compernolle Imec-MICT, Ghent University, Ghent, Belgium

Grazia Concilio Department of Architecture and Urban Studies, Politecnico di Milano, Milan, Italy

Petter Falk Research Institutes of Sweden, Karlstad University, Karlstad, Sweden

Giovanni Lanza Department of Architecture and Urban Studies, Politecnico di Milano, Milan, Italy

Francesco Molinari Department of Architecture and Urban Studies, Politecnico di Milano, Milan, Italy

Paola Pucci Department of Architecture and Urban Studies, Politecnico di Milano, Milan, Italy

Lieven Raes Digitaal Vlaanderen, Brussels, Belgium

Paolo Raineri Como, Italy

Bart Rosseau Dienst Data En Informatie, Bedrijfsvoering, Stad Gent, Belgium

Matteo Satta Issy Média, Issy-Les-Moulineaux, France

Jonas Verstraete Dienst Data En Informatie, Bedrijfsvoering, Stad Gent, Belgium

Nils Walravens Imec-SMIT, Vrije Universiteit Brussel, Brussels, Belgium

Part I
The Data Shake: Open Questions and Challenges for Policy Making

Chapter 1
The Data Shake: An Opportunity for Experiment-Driven Policy Making

Grazia Concilio and Paola Pucci

Abstract The wider availability of data and the growing technological advancements in data collection, management, and analysis introduce unprecedented opportunities, as well as complexity in policy making. This condition questions the very basis of the policy making process towards new interpretative models. Growing data availability, in fact, increasingly affects the way we analyse urban problems and make decisions for cities: data are a promising resource for more effective decisions, as well as for better interacting with the context where decisions are implemented. By dealing with the operative implications in the use of a growing amount of available data in policy making processes, this contribution starts discussing the chance offered by data in the design, implementation, and evaluation of a planning policy, with a critical review of the evidence-based policy making approaches; then it introduces the relevance of data in the policy design experiments and the conditions for its uses.

Keywords Policy experiments · Learning cycles · Evidence-based policy making · Policy cycle

1.1 Introduction

The wider availability of data and the growing technological advancements in data collection, management, and analysis introduce unprecedented opportunities, as well as complexity in policy making. This condition questions the very basis of the policy making process towards new interpretative models.

Growing data availability, in fact, increasingly affects the way we analyse urban problems and make decisions for cities: data are a promising resource for more

G. Concilio (✉) · P. Pucci
Department of Architecture and Urban Studies, Politecnico di Milano, Milan, Italy
e-mail: grazia.concilio@polimi.it

P. Pucci
e-mail: paola.pucci@polimi.it

© The Author(s) 2021
G. Concilio et al. (eds.), *The Data Shake*,
PoliMI SpringerBriefs,
https://doi.org/10.1007/978-3-030-63693-7_1

effective decisions, as well as for better interacting with the context where decisions are implemented.

Such multiplicity of data and its different sources poses several challenges to policy making. First, the availability of a large amount of data improves the accuracy and completeness of the measurements to capture phenomena that were previously difficult to investigate but, at the same time, increases the level of complexity in the approaches finalized to process, integrate, and analyse this data (Einav and Levin 2013).

Second, processing data is not neutral and irrelevant for its usability in decision making processes. The selection and interpretation of a large amount of unstructured information, deriving from data, requires a human based approach finalized to find what emerging correlations between data are significant or not. In doing so, tools to examine data are crucial, considering that non-human agents develop potentially partial ways of understanding the world around them (Mattern 2017) and that some tools, such as algorithms, can act as technical counters to liberty (Greenfield 2017, p. 257).

Third, the huge amount of real-time, automated, volunteered data pushes towards an epistemological change in the methodological approaches of empirical sciences, transforming how we observe and interpret urban phenomena, moving from a "*hypothetical-deductive method, driven by an incremental process of falsification of previous hypotheses*" to "*an inductive analysis at a scale never before possible*" (Rabari and Storper 2015, p. 33). In addition to using data to test previous hypotheses, new phenomena and correlations between them may emerge as the result of the massive processing of data (Kitchin 2014), with repercussions in decision making activities in a short-medium term planning perspective.

Finally, while data is a non-neutral tool for addressing planning issues, the actors that produce, manage and own data, both public and private—with the latter typically being corporations active in fields outside traditional regulations—configure an unprecedented geography of power, a more complex arena in which urban problems are defined, discussed and finally addressed by new constellations of actors.

These different implications and conditions related to the larger availability of data, from data production, management, and analytics, to its potential in decision making processes for both private and public actors, find synthesis in the expression of "data shake".

By dealing with the operative implications in the use of a growing amount of available data in policy making processes, this article discusses the chance offered by data in the design, implementation, and evaluation of a planning policy, starting from a critical review of the evidence-based policy making approaches (Sect. 1.2), for introducing the relevance of data in the policy design experiments (Sect. 1.3) and the conditions for its uses. Acknowledging the impossibility of simply relying on data for framing urban issues and possible solutions to them, and considering the potential disruptions brought by data into the urban planning practices, this paper focuses on policy processes where data is used, rather than simply focusing on technological solutions fostered by data.

1.2 Evidence-Based Policy Making: New Chances Coming from the Data Shake

1.2.1 About Evidence-Based Policy Making

Evidence-based policy making (EBPM) represents an effort started some decades ago to innovate and reform the policy processes for the sake of more reliable decisions; the concept considers evidence as being a key reference for prioritizing adopted decision criteria (Lomas and Brown 2009; Nutley et al. 2007; Pawson 2006; Sanderson 2006). The key idea is to avoid, or at least reduce, policy failures rooted in the ideological dimension of the policy process, by adopting a rationality having a solid scientific basis. The fact that evidence should come from scientific experts and guide the policy makers' activities appeared and still appears a panacea to several scientists in the policy making and analysis domain: this makes evidence based policy making a sort of expectation against which policy makers, and political actors in general, can be judged (Parkhurst 2017, p. 4).

The evidence-based policy movement, as Howlett (2009) defines it, is just one effort among several others to be undertaken by governments to enhance the efficiency and effectiveness of public policy making. In these efforts, it is expected that, through a process of theoretically informed empirical analysis, governments can better learn from experience, avoid errors, and reduce policy-related contestations.

Finding a clear definition of the concept is not easy. In the policy literature, the meaning is considered sort of "self-explaining" (Marston and Watts 2003) and is associated with empirical research findings. Many scholars refer the evidence-based policy concept as evolving from the inspiring experience in medicine: here, research findings are key references for clinical decisions, and evidence is developed according to the so-called "golden standard" of evidence gathering that is the "randomized controlled trial" a comparative approach to assess treatments against placebos (Trinder 2000). Following the large importance assumed in medicine and healthcare, there was then an increase in research and policy activists pushing for evidence-based approach in other domains of policy making more related to social sciences and evidence produced by the social science research, covering a wider range of governmental decision making processes (Parsons 2001).

Moreover, the spreading of the evidence-based concept in policy making corresponds to the infiltration of instrumentalism in public administration practices following the managerial reforms of the last decades: the key value assigned to effectiveness and efficiency by managerialism represented a driving force for evidence-based policies (Trinder 2000, p. 19), so emphasizing procedures, sometimes at the expense of substance.

The key discussion is on what makes evidence such: the evidence-based approach in policy making is strictly correlated to the procedure, empirical procedure, that makes evidence reliable. The spreading of the concept made social sciences look at their procedural and methodological approaches to collect evidence although

Table 1.1 Key concerns raised about the emphasis on evidence in policy making

Key critics to EBPM	References (quoted by Howlett)
Evidence is only one factor involved in policy making and is not necessarily able to overcome other	Davies (2004); Radin and Boase (2000); Young et al. (2002)
Data collection and analytical techniques employed in its gathering and analysis by specially trained policy technicians may not be necessarily superior to the experiential judgments of politicians and other key policy decision makers	Jackson (2007); Majone (1989)
The kinds of "high-quality" and universally acknowledged evidence essential to "evidence-based policy making" often has no analogue in many policy sectors, where generating evidence using the "gold standard" of random clinical trial methodologies may not be possible	Innvaer et al. (2002); Pawson et al. (2005)
Government efforts in this area may have adverse consequences both for themselves in terms of requiring greater expenditures on analytical activities at the expense of operational ones	Hammersley (2005); Laforest and Orsini (2005)

Source Howlett (2009, p. 155)

the research categories of social science are missing deeply structured empirical approaches.

"*Evidence matters for public policy making*" as Parkhurst (2017, p. 3) underlines by presenting and discussing three examples,[1] despite the concept collecting several critics and concerns all together incriminating the supporters of the evidence-based concept of being scarcely aware of the socio-political complexity of policy making processes. Howlett (2009) has summarized such critics and concerns in four main categories (Table 1.1).

Public policy issues have a prevailing contested, socio political nature that amplifies the complexity of evidence creation processes: decision processes in public policy making is not a standard, not a rational decision exercise; it is more a "*struggle over ideas and values*" (Russell et al. 2008, p. 40, quoted by Parkhurst 2017, p. 5), it is related to visions of the future and principles, so hardly manageable through rational approaches and science.[2] In this respect, Parsons (2001) considers that, when values are involved more than facts and evidence, policy processes are required which are

---

[1] Among them: the risk of SIDS (sudden infant death syndrome) for front slipper children (2005), the research done by the oil company Exxon on the effects of fossil fuels on the environment (1970–1980); the security risk posed by Iraqi regime in 2003, according to the US President George W. Bush (p. 3).

[2] A widely discussed example in literature is related to policy making on abortion in different countries: debates on abortion were more related to women rights against the rights of unborn as

"more democratic and which can facilitate … deliberation and public learning"
(p. 104).

1.2.2 Evidence-Based Policy Making and the Data Shake: The Chance for Learning

The increasing production of huge amounts of data, its growing availability to different political subjects, and the wide exploration of the data potentials in decision making for both private and public actors, are proceeding in parallel with the fast advancements in technologies for data production, management and analytics. This is what we call the data shake, and it is not only related to the larger and larger availability of data but also to the faster and faster availability of data-related technologies. As a consequence, data-driven approaches are being applied to several diverse policy sectors: from health to transport policy, from immigration to environmental policy, from industrial to agricultural policy. This is shaking many domains and, as never before, also the social science domain: the larger availability of data, in fact, and easy to use data-related technologies, make data usable also by non-experts so widening the complexity of social phenomena.

Nevertheless, although the data shake appears to have promising and positive consequences in policy and policy making, existing literature underlines the role of some consolidated critical factors affecting the chance for data to achieve such a promising perspective. As highlighted by Androutsopoulou and Charalabidis (2018, p. 576), one of the key factors is *"the demand for broader and more constructive knowledge sharing between public organisations and other societal stakeholders (private sector organisations, social enterprises, civil society organisations, citizens)."* Policy issues *"require negotiation and discourse among multiple stakeholders with heterogeneous views, tools that allow easy data sharing and rapid knowledge flows among organisations and individuals have the potential to manage knowledge facilitating collaboration and convergence"*. The response to such a demand implies relevant expertise in organizations to adopt the "right" data, among the wide range of available data sets, to analyse the data and to produce the effective evidence to guarantee knowledge production and sharing.

Another key factor is related to the use of data when dealing with social problems: as again highlighted by Androutsopoulou and Charalabidis (2018), there is an issue of proper use of data to develop a reliable description of the problem and the formulation of effective policy measures. Also in this case, the selection of the proper data set or sets, the application of a data integration strategy, the design of analytical tools or models able to be effective in representation without losing the richness of information embedded in data, and the consequent formulation of effective policy measure consistent with the problem description are not simple rational decisions

---

well as to what a good society should look like; none of this debate can be definitely be closed with science or scientific evidence.

and imply also the consideration of approaches to public debates to negotiate both the vision and interpretation of the social problem and the solution to adopt.

The simple existence of more and different data and the related availability of technical tools do not grant the solution of the issues identified by the opponents to the evidence-based policy making concept. This last point explains why Cairney (2017, pp. 7,8,9) concludes that attention is needed to the politics of evidence-based policy making: scientific technology and methods to gather information *"have only increased our ability to reduce but not eradicate uncertainty about the details of a problem. They do not remove ambiguity, which describes the ways in which people understand problems in the first place, then seek information to help them understand them further and seek to solve them. Nor do they reduce the need to meet important principles in politics, such as to sell or justify policies to the public (to respond to democratic elections) and address the fact that there are many venues of policy making at multiple levels (partly to uphold a principled commitment, in many political systems, to devolve or share power)"*.

Better evidence, possibly available thanks to the data shake, may eventually prove that a decision is needed on a specific issue, or prove the existence of the issue itself; still it cannot yet clarify whether the issue is the first in priority to be considered or show what the needed decision is: the uncertainty and unpredictability of socio-political processes remain unsolved although better manageable.

Still, something relevant is available out there. Although the socio-political complexity of policy making stays unchanged, the data shake is offering an unprecedented chance: the continuous production of data throughout the policy making process (design, implementation, and evaluation) creates the chance to learn through (not only for neither from) the policy making process. This opportunity is concrete as never before. The wide diversity of data sources, their fast and targeted production, the available technologies that produce easy to use analytics and visualizations create the chance for a shift from *learning for/from policy making* into *learning by policy making* so allowing the improvement of the substance and procedure at the same time as a continuous process.

The learning opportunity is directly embedded in the policy making process as the chance to shape social behaviours, responses, and achieve timely (perhaps even real-time) effects (Dunleavy 2016) is out there. Learning by (doing in) policy making is possible and benefits from a new role of evidence: no longer (or not only) a way to legitimate policy decisions, no longer (or not only) an expert guide to more effective and necessary policy making rather a means for learning, for transforming policy making into a collective learning process. This is possible as the data shake gives value to the evidence used over time (Parkhurst 2017) so enabling its experimental dimension.

1.3 The Smart Revolution of Data-Driven Policy Making: The Experimental Perspective

1.3.1 About Policy Experiments and Learning Cycles

In social science, a policy experiment is any "[…] *policy intervention that offers innovative responses to social needs, implemented on a small scale and in conditions that enable their impact to be measured, prior to being repeated on a larger scale, if the results prove convincing*" (European Parliament and Council 2013, art.2 (6)). Policy experiments form a useful policy tool to manage complex long-term policy issues by creating the conditions for "ex-ante evaluation of policies" (Nair and Howlett 2015): learning from policy experimentation is a promising way to approach "wicked problems" which are characterised by knowledge gaps and contested understandings of future (McFadgen and Huitema 2017); experiments carried out in this perspective, in fact, generate learning outcomes mainly made of relevant information for policy and under dynamic conditions (McFadgen 2013).

The concept of policy experimentation is not new. An explanatory reconstruction of the concept development has been carried out by van der Heijden (2014), who quoted John Dewey (1991 [1927]) and Donald Campbell (1969, p. 409) as seminal contributions to it. In particular: Dewey already considered that policies should "*be treated as working hypotheses, not as programs to be rigidly adhered to and executed. They will be experimental in the sense that they will be entertained subject to constant and well-equipped observation of the consequences they entail when acted upon, and subject to ready and flexible revision in the light of observed consequences*" (pp. 202–203); while Campbell considered experimental an approach in which new programs are tried out, as they are conceived in a way that it is possible both to learn whether they are effective and to imitate, modify, or discard them on the basis of apparent effectiveness on the multiple imperfect criteria available (p. 409). van der Heijden considers that Dewey and Campbell had in mind the idea of experimenting with the content of policy programs (testing, piloting, or demonstrating a particular policy design), rather than the process of policy design.

Still, as van der Heijden observes, silent remains as to the actual outcomes of such experimentations, and this consideration makes the scope of his article that develops two main conclusions:

- experimentation in environmental policy is likely to be successful if participation comes at low financial risk and preferably with financial gain (see Baron and Diermeier 2007; Croci 2005, quoted by van der Heijden);
- in achieving policy outcomes, the content of the policy-design experiments matters more than the process of experimentation.

Intercepting both policy contents and experimentation process, and focussing on the governance design of policy making, McFadgen and Huitema (2017) identified three types of experiments: the expert driven "technocratic" model, the participatory

Fig. 1.1 Learning effects in policy experimentation (extracted from Table 1 in McFagden and Huitema 2017, pp. 3–22)

Learning effect	Definition
Cognitive	- knowledge acquisition; - improved structuring of existing knowledge
Normative	- change perspectives - goal convergence
Relational	- increase in understanding of others' mind-sets; - increase trust and cooperation

"boundary" model, and the political "advocacy" model. These models differ in their governance design and highlight how experiments produce learning; together with what types of learning they activate.

In the technocratic model, experts work as consultants; they are asked to produce evidence to support or refute a claim within the context of political disagreement. In this model, policy makers are out of the experiment, but they supply in advance the policy problem and the solution to be tested.

In the boundary model, experiments (working on borders among different points of view) have a double role: producing evidence but also debating norms and developing a common understanding. In this kind of experiment, the involvement of different actors is crucial for the experiment to be productive of knowledge and discussion at different cognitive levels (practical, scientific, political).

In the advocacy model the experiment is aimed at reducing objections to a predefined decision. These experiments are tactical and entirely governed by policy makers who are obviously interested in involving other actors. This kind of experiment can also be initiated by non-public actors, even with different scopes.

McFadgen and Huitema (2017) also highlight the different learning taking places during the three different experimental models. They distinguish mainly three kinds of learning (Fig. 1.1).

Taking into consideration the goals and the differences in participants of the three experiment models, McFadgen and Huitema (2017) find that: technocratic experiments mainly generate high levels of cognitive learning, little normative, and some relational learning, which is mainly due to the disconnection between experiments and the policy makers; boundary experiments are expected to produce relational and normative learning while low levels of cognitive learning due to the large importance assigned to debating and sharing; advocacy experiments cognitive and normative learning are expected to be activated but little relational learning and this is due to the intentional selection of participants.

Learning in policy experiments is crucial and is mainly related to the opportunity embedded in learning to become appropriation of the knowledge developed throughout the experiment. Consequently, the rationale behind an experimental approach to policy making is to boost public policy makers' ownership and commitment, thus possibly increasing the chances that successful experiments are streamlined into public policy.

The experimental dimension, especially in the boundary and the advocacy models, is crucial in policy design and policy implementation. It makes the policy evaluation scope transversal to the other steps of the policy cycle—described by Verstraete et al. (2021)—as well as supportive of the other steps. It transforms policy making into an experimental process as it introduces co-design and co-experience paving the way for embedding new points of view and new values in the context of the policy. Design and implementation, in this perspective, become reciprocal and integrated (Concilio and Celino 2012; Concilio and Rizzo 2012) and:

- learning is enhanced and extended to participants by designing "with", not merely "for";
- exchange and sharing of experiences are more effective than information transfer and sharing;
- involved actors become the owners of the socio-technical solutions together with technological actors and decision makers;
- changes in behaviours (the main goal of any policy making) are activated throughout the experiments.

Based on this, different levels of integration are possible and, among them, the most advanced is the so-called triple-loop learning flow in policy experimentations (Yuthas et al. 2004; see also Deliverable 3.1 by the Polivisu Project[3]).

1.3.2 Policy Cycle Model Under Experimental Dimension

As introduced in the previous section, the experimentations and the reflection on the operative implications in the use of data in urban management and decision making processes are at the base of a consistent production of critical ex-post evaluations on the potentials and limits of data-informed policy making produced in the last years (e.g. Poel et al. 2015; Lim et al. 2018).

The process of policy creation has been left in the background by the focus on the content, rather than the process of policy design and, in some cases, without a proper reflection about the selection, processing, and use of data to identify individual or collective human needs and formulate solutions that "can be not arrived at algorithmically" (…); and which cannot be "encoded in public policy, without distortion" (Greenfield 2017, p. 56).

Actually, it is well accepted that a policy process is not a linear and deterministic process; it is a set of decisions and activities that are linked to the solution of a collective problem where the "connection of intentionally consistent decisions and activities taken from different public actors, and sometimes private ones (are addressed) to solve in a targeted way a problem" (Knoepfel et al. 2011, p. 29).

In this process, data offer support for strategic activities by aggregating information on a time series that support and validate prediction models for long-term planning; for tactical decisions, conceived as the evidence-informed actions that are needed to implement strategic decisions and, finally, for operational decisions, giving

---

[3] https://www.polivisu.eu/public-deliverables.

support to day-to-day decision making activities in a short-term planning perspective (Semanjski et al. 2016).

From a policy perspective, strategic, tactical, and operational decisions use, and are supported by, data in different ways along the stages of a policy making process. In the design, in the implementation, and in the evaluation of a policy, data provides insights in allowing the possibility to discover trends and to eventually analyze their developing explanation; in fostering public engagement and civic participation; in dynamic resource management; and, finally, in sustaining the development of "*robust approaches for urban planning, service delivery, policy evaluation and reform and also for the infrastructure and urban design decisions*" (Thakuriah et al. 2017, p. 23). Among them, an approach in which data may support a policy making process dealing with a different time frame and multi-actor perspectives can be based on the policy cycle model, which means conceiving policy as a process composed of different steps (Marsden and Reardon 2017) to which data contributes differently.

The policy cycle, here not be interpreted as a rigorous, formalistic guide to the policy process, but as an idealized process, as a "*means of thinking about the sectoral realities of public policy processes*", has the potential to capture the potential of data shake if used in a descriptive way more than in normative one.

This policy model can be conceptualized as a data-assisted policy experimentation cycle, consisting of interrelated cyclical stages: the stages are strongly interdependent, integrated, and overlapping due to the broad availability of data at the core of policy making's experimental dimension.

In doing so, the policy cycle model can represent a "bridge", a sort of "boundary object" (Star and Griesemer 1989) in which different operational and disciplinary dimensions (planning, data analytics, data mining) can interact and cross-fertilize each other since it offers an organized structure, in which data provides a viable base for acting in each stage.

Based on this, the major weaknesses recognized in the policy cycle model, considered too simplistic in practice, giving a false impression of linearity and discrediting its assumption of policy as sequential in nature[4] (Dorey 2005; Hill 2009; Howlett and Ramesh 2003; Ryan 1996) may be overpass thanks to the experimental perspective, able to foster a less linear interpretation of policy cycle, transformed in a continuous process in which overlap among policy stages.

---

[4] Among the critical arguments: the inability of this model to explain what causes policies to advance from one stage to another, the predetermined manner in with each stage in the cycle occurring in a precise, far from actual fact (Howlett and Ramesh 2003), because policy needs to be designed and continuously revised to take into account external conditions and adapt to their eventual change. Their effects are often indirect, diffuse, and take time to appear; policy making depends on politics, people, socio-economic factors, and other previous and ongoing policies.

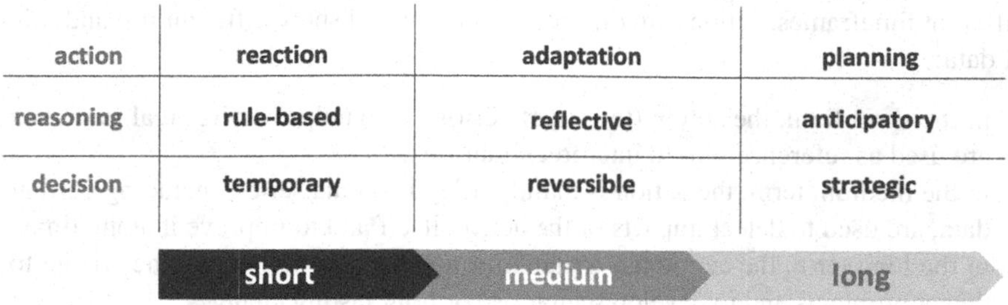

action	reaction	adaptation	planning
reasoning	rule-based	reflective	anticipatory
decision	temporary	reversible	strategic

short	medium	long

Fig. 1.2 Decision/reasoning along diverse timeframes

1.3.3 The Time Perspective in the Experimental Dimension of Policy Making[5]

Decisions for and about cities are made at different urban scales, refer to different strategic levels and have different time perspectives, with reciprocal interdependencies that are changing due to data availability. Here we mainly focus on the interplay between the different steps of decisions in policy making (those introducing the long-time perspective) and those necessary for the daily management of the city (connected to the shortest, real-time perspective), an intersection at which data can play a key role (Fig. 1.2).

Short-term management is embedded in the smart sphere of decisions impacting cities: here decisions are less analytic and more routine. Routines may depend on data-driven learning mechanisms (also using data series) supporting smart systems to recognize situations and apply solutions and decisions that have already been proven to work. The decision has a temporary value related to the specific conditions detected in a precise moment by the smart system.

Opposite to real-time decisions, policy making works in a long-time perspective. Anticipatory is the prevailing mode for reasoning in this case data-driven models are often adopted as supporting means to deal with the impacts of the policy measures, representing thus a relevant source for exploring decision options mainly having a strategic nature (since they consider recurring issues and aim at more systemic changes).

Between short-term and long-term decisions a variety of situations is possible, which may be considered as characterized by decisions having a reversible nature: they are neither strategic in value (like those oriented to a long term perspective for systemic changes), nor aiming at dealing with temporary, contingency situations asking decisions which are known as having the same (short) duration of the phenomenon to be managed. For such decisions, the reasoning is not (fully) anticipatory and its temporariness allows reflection as embedded in action. Within the three

---

[5]This paragraph is belonging to a recent publication: Concilio G, Pucci P, Vecchio G, Lanza G (2019) Big data and policy making: between real time management and the experimental dimension of policies. In: Misra S. et al. (eds) Computational science and its applications—ICCSA 2019. Lecture Notes in Computer Science, vol 11620, Springer, Cham, pp 191–202.

different timeframes, actions are different in nature and show different use and role of data:

- in the short term, the action (the smart action) is mainly reactive; real time data are used as reference info to interpret situations;
- in the medium term, the action is mainly adaptive; data series, including current data, are used to detect impacts of the action itself and to improve it along time;
- in the long term, the action has a planning nature; data series become crucial to detect problems and to develop scenarios for long lasting changes.

The interdependency between policy design, implementation and evaluation is strictly related to two factors, especially when considering the role (big) data can play. Design and implementation can be clearly and sequentially distinguished when a systematic, impact-oriented analysis is possible at the stage of design as it allows a clear costs and benefits assessment of different action opportunities (Mintzberg 1973).

Comprehensive analyses have the value to drive long range, strategic actions, and consequently show a clear dependency and distinction of the implementation from the design cycle. At the same time, the bigger is the uncertainty (not only related to the possible lack of data, rather also consequent to the high complexity of the problem/phenomenon to be handled), the smaller is the chance to carry out a comprehensive analysis.

Therefore, goals and objectives cannot be defined clearly and the policy making shifts from planning towards an adaptive mode. Inevitably, this shift reduces the distance between design and implementation, transforming policy design into a more experimental activity that uses learning from implementation into food for design within adaptive dynamics (Fig. 1.3).

Coherently with the discussion on the time frame perspective adopted, it may be clear that a merge between policy design and implementation is consistent with

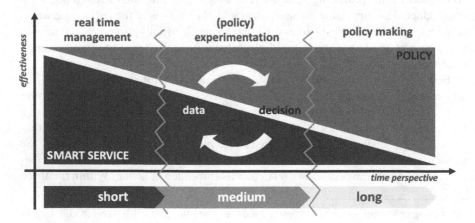

Fig. 1.3 Real time management vs policy making

the situation described in the medium term: within an adaptive mode for decisions, policy making can clearly become experimental.

1.4 Conclusions: Beyond the Evidence-Based Model

Evidence based policy making is surely the key conceptual reference when trying to grasp the potentials that the growing availability of data and related technologies offer to policy making. As it is clear from the previous paragraphs, the concept has been widely discussed in the literature and can be considered the key antecedent of experiment-driven policy making.

Experiments may refer to both the policy strategies and measures. They can reduce the risk of trial-errors approaches while considering the learning in action opportunity to improve, adapt, adjust the policy while making it in order to increase its capacity to affect the context in an evolving manner.

Differently from Mintzberg's considerations (1973), the merge between policy design and implementation does not represent a sort of inevitable, but not preferred option when a comprehensive analysis is not possible. In the era of data availability, this merge can be looked at as an opportunity to create policies while verifying the policies themselves throughout their interactions with the contexts.

The growing availability of diverse and rich data sets represents an opportunity for evidence to be transformed into a more valuable resource then what it was intended to be by the evidence-based policy making supporters: not only, or not necessarily, a means to support the scientific rationality of the decision making process, rather the drivers to reflection and learning through action.

References

Androutsopoulou A, Charalabidis Y (2018) A framework for evidence based policy making combining big data, dynamic modelling and machine intelligence. In: Kankanhalli A, Ojo A, Soares D (eds) Proceedings of the 11th international conference on theory and practice of electronic governance, Galway, Ireland, 4–6 April 2018, pp 575–583

Baron DP, Diermeier D (2007) Strategic activism and nonmarket strategy. J Econ Manag Strat 16:599–634

Cairney P (2017) The politics of evidence-based policy making. In: Oxford research encyclopedia of politics. Available at http://eprints.lse.ac.uk/68604/1/Parkhurst_The%20Politics%20of%20Evidence.pdf. Accessed on June 2020

Campbell D (1969) Reforms as experiments. Am Psychol 24(4):409–429

Concilio G, Celino A (2012) Learning and innovation in living Lab environments. In: Schiuma G, Spender JC, Yigitcanlar T (eds) Proceedings of the international forum on knowledge assets dynamics—knowledge, innovation and sustainability: integrating micro and macro perspective, Matera, Italy, 13–15 June 2012

Concilio G, Rizzo F (2012) Enabling situated open and participatory design processes by exploiting a digital platform for open innovation in smart cities. In: Miettinen S, Valtonen A (eds) Service design with theory, Lapland University Press, pp 66–72

Croci E (2005) The handbook of environmental voluntary agreements. Springer, Dordrecht

Davies P (2004) Is evidence-based government possible? Jerry Lee Lecture presented at the 4th annual campbell collaboration colloquium, Washington, DC. 19 February 2004

Dewey J (1991 [1927]) The public and its problems. Swallow Press, Ohio

Dorey P (2005) Policy making in Britain: an introduction. Sage, London, UK

Dunleavy P (2016) 'Big data' and policy learning. In: Stoker G and Evans M (eds) Evidence-based policymaking in social science. Methods that matter. The Policy Press, Bristol, pp 143–168

EC (European Commission) (2013) Guide to social innovation. Brussels, European Commission—DG Regional and Urban Policy

Einav L, Levin JD (2013) The data revolution and economic analysis. NBER Working Paper Series, 19035

Greenfield A (2017) Radical technologies: the design of everyday life. Verso, Brooklyn

Kitchin R (2014) Big data, new epistemologies and paradigm shifts. Big Data Soc 1(1)

Knoepfel P, Larrue C, Varone F, Hill M (2011) Public policy analysis. The Policy Press, Bristol

Hammersley M (2005) Is the evidence-based practice movement doing more good than harm? Reflections on Iain Chalmers' case for research-based policy making and practice. Evid Policy 1(1):85–100

Hill M (2009) The public policy process. Pearson, London

Howlett M (2009) Policy analytical capacity and evidence-based policy-making: lessons from Canada. Can Public Adm / Administration publique du Canada 52(2) (june/juin 2009):153–175

Howlett M, Ramesh M (2003) Studying public policy: policy cycles and Policy sub-systems. Oxford University Press, Oxford

Innvaer S, Vist G, Trommald M, Oxman A (2002) Health policy-makers' perceptions of their use of evidence: a systematic review. J Health Serv Res Policy 7(4):239–245

Jackson PM (2007) Making sense of policy advice. Public Money Manag 27(4):257–264

Laforest R, Orsini M (2005) Evidence-based engagement in the voluntary sector: lessons from Canada. Soc Policy Adm 39(5):481–497

Lim C, Kim KJ, Maglio PP (2018) Smart cities with big data: reference models, challenges and considerations. Cities 82:86–99

Lomas J, Brown A (2009) Research and advice giving: a functional view of evidence informed policy advice in a Canadian ministry of health. Milbank Q 87(4):903–926

Majone G (1989) Evidence, argument, and persuasion in the policy process. Yale University Press, New Haven, CO

Marsden G, Reardon L (2017) Questions of governance: rethinking the study of transportation policy. Transp Res Part A: Policy Pract 101 238–251. https://doi.org/10.1016/j.tra.2017.05.008

Marston G, Watts R (2003) Tampering with evidence: a critical appraisal of evidence-based policy making. Draw Board: Aust Rev Public Aff 3(3):143–163

Mattern S (2017) Mapping's intelligent agents. Places J. https://doi.org/10.22269/170926

McFadgen B (2013) Learning from policy experiments in adaptation governance. Paper presented at the 1st international conference on public policy. Grenoble, France, 26–28 June 2013

McFadgen B, Huitema D (2017) Stimulating learning through policy experimentation: a multi-case analysis of how design influences policy learning outcomes in experiments for climate adaptation. Water 9(9):648

Mintzberg H (1973) Strategy-making in three modes. Calif Manag Rev 16(2):44–53

Nair S, Howlett M (2015) Scaling up of policy experiments and pilots: a qualitative comparative analysis and lessons for the water sector. Water Resour Manag 29:4945–4961

Nutley S M, Walter I, Davies HTO (2007) Using evidence: how research can inform public services. Policy Press, Bristol

Parkhurst J (2017) The politics of evidence. from evidence-based policy to the good governance of evidence. Routledge, Abingdon

Parsons W (2001) Modernising policy-making for the twenty first century: the professional model. Public Policy Adm 16(3):93–110

Pawson R (2006) Evidence-based policy: a realist perspective. Sage, London

Pawson R, Greenhalgh T, Harvey G, Walshe K (2005) Realist review—a new method of systematic review designed for complex policy interventions. J Health Serv Res Policy 10 (Supplement 1):21–34

Poel M, Schroeder R, Treperman J, Rubinstein M, Meyer E, Mahieu B, Scholten C, Svetachova M (2015) Data for policy: a study of big data and other innovative data-driven approaches for evidence-informed policymaking. Report about the state-of-the-art. Joint venture between Technopolis group, Oxford internet Institute and the Centre for European Studies

Pucci P, Vecchio G, Concilio G (2018) Big data and urban mobility: a policy making perspective. Transportation Research Procedia 2352–1465, Elsevier B.V. World Conference on Transport Research, Mumbai, India, 26–31 May 2019

Rabari C, Storper M(2015) The digital skin of cities: urban theory and research in the age of the sensored and metered city, ubiquitous computing and big data. Cambridge Journal of Regions. Econ Soc 8(1):27–42. https://doi.org/10.1093/cjres/rsu02

Radin BA, Boase JP (2000) Federalism, political structure, and public policy in the United States and Canada. J Comp Policy Anal 2(1):65–90

Russell J, Greenhalgh T, Byrne E, Mcdonnell J (2008) Recognizing rhetoric in health care policy analysis. J Health Serv Res & Policy 13(1):40–46

Ryan N (1996) Some advantages of an integrated approach to implementation analysis: a study of the Australian industrial policy. Public Adm 74(4):737–753

Sanderson I (2006) Complexity, 'practical rationality' and evidence-based policy making. Policy Polit 34 (1):115–1322

Semanjski I, Bellens R, Gautama S, Witlox F (2016) Integrating big data into a sustainable mobility 2.0 planning support system. Sustainability, 8, 1142

Star S, Griesemer J (1989) Institutional ecology, 'translations' and boundary objects: amateurs and professionals in Berkeley's museum of vertebrate zoology, 1907–39. Soc Stud Sci 19(3):387–420

Thakuriah P, Tilahun NY, Zellner M (2017) Big data and urban informatics: innovations and challenges to urban planning and knowledge discovery. In: P. Thakuriah et al. (eds) Seeing cities through big data. Springer Geography

Trinder L (2000) Introduction: the context of evidence-based practice. In: Trinder L, Reynolds S (eds) Evidence-based practice: a critical appraisal. Blackwell Science, Oxford

van der Heijden J (2014) Experimentation in policy design: insights from the building sector. Policy Sci 47:249–266

Verstraete J, Acar F, Concilio G, Pucci P (2021) Turning data into actionable policy insights. In: G Concilio, P Pucci, L Raes, G Mareels (eds) The data shake. opportunities and obstacles for urban policy making. Springer, PolimiSpringerBrief

Young K, Ashby D, Boaz A, Grayson L (2002) Social science and the evidence-based policy movement. Soc Policy Soc 1(3):215–224

Yuthas K, Dillard J, Rogers R (2004) Beyond agency and structure: triple-loop learning. J Bus Ethics 51(2):229–243

Grazia Concilio Associate professor in Urban Planning and Design at DAStU, Politecnico di Milano. She is an engineer; PhD in "Economic evaluation for Sustainability" from the University of Naples Federico II. She carried out research activities at the RWTH in Aachen, Germany (1995), at IIASA in Laxenburg, Austria (1998) and at the Concordia University of Montreal, Canada, (2002); she is reviewer for several international journals and former member (in charge of assessment task of LL new applications) of ENoLL (European Network of open Living Lab). Team member in several research projects; responsible for a CNR research program (2001) and coordinator of a project funded by the Puglia Regional Operative Programme (2007-2008) and aiming at developing an e-governance platform for the management of Natural Parks.

She has been responsible on the behalf POLIMI of the projects Peripheria (FP7), MyNeigh-bourhood|MyCity (FP7), Open4Citizens (Horizon 2020 www.open4citizens.eu); she is currently responsible for the Polimi team for the projects Designscapes (Horizon 2020 www.design scapes.eu), Polivisu (Horizon 2020 www.polivisu.eu) together with Paola Pucci, and MESOC (Horizon 2020 www.mesoc-project.eu). She is coordinating the EASYRIGHTS project (Horizon 2020 www.easyrigths.eu). She is the author of several national and international publications.

Paola Pucci Full Professor in Urban planning, she has been Research Director of the PhD course in Urban Planning Design and Policy (2012–2018) at the Politecnico di Milano. From 2010 to 2011 she taught at the Institut d'Urbanisme in Grenoble Université Pierre Méndes France at Bachelor, Master and PhD levels and currently visiting professor at European universities. She has taken part, also with roles of team coordinator, in national and international research projects funded on the basis of a competitive call, dealing with the following research topics: Mobility policy and transport planning, mobile phone data and territorial transformations and including EU ERA-NET project "EX-TRA – EXperimenting with city streets to TRAnsform urban mobility"; H2020 - SC6-CO-CREATION-2016-2017 "Policy Development based on Advanced Geospatial Data Analytics and Visualisation", EU Espon Project, PUCA (Plan, urbanisme, architecture) and PREDIT projects financed by the Ministère de l'Ecologie, du Développement et de l'Aménagement durable (France). She has supervised and refereed different graduate, postgraduate and PhD theses at Politecnico di Milano, Université Paris Est Val de Marne, Ecole Superieure d'Architecture de Marseille, Université de Tours. She has been Member of the evaluation panel for the Netherlands Organisation for Scientific Research (NWO, 2017), and Member of the NEFD Policy Demonstrators commissioning panel for the ESRC _Economic and Social Research Council. Shaping Society (Uk), on the topic "New and Emerging Forms of Data - Policy Demonstrator Projects (2017).

Chapter 2
Data Ownership and Open Data: The Potential for Data-Driven Policy Making

Nils Walravens, Pieter Ballon, Mathias Van Compernolle, and Koen Borghys

Abstract As part of the rhetoric surrounding the Smart City concept, cities are increasingly facing challenges related to data (management, governance, processing, storage, publishing etc.). The growing power acquired by the data market and the great relevance assigned to data ownership rather than to data-exploitation knowhow is affecting the development of a data culture and is slowing down the embedding of data-related expertise inside public administrations. Concurrently, policies call for more open data to foster service innovation and government transparency. What are the consequences of these phenomena when imagining the potential for policy making consequent to the growing data quantity and availability? Which strategic challenges and decisions do public authorities face in this regard? What are valuable approaches to arm public administrations in this "war on data"? The Smart Flanders program was initiated by the Flemish Government (Belgium) in 2017 to research and support cities with defining and implementing a common open data policy. As part of the program, a "maturity check" was performed, evaluating the cities on several quantitative and qualitative parameters. This exercise laid to bare some challenges in the field of open data and led to a checklist that cities can employ to begin tackling them, as well as a set of model clauses to be used in the procurement of new technologies.

Keywords Data ownership · Government transparency · Data policy

N. Walravens (✉) · P. Ballon · K. Borghys
Imec-SMIT, Vrije Universiteit Brussel, Brussels, Belgium
e-mail: nils.walravens@imec.be

P. Ballon
e-mail: pieter.ballon@imec.be

K. Borghys
e-mail: koen.borghys@imec.be

M. Van Compernolle
Imec-MICT, Ghent University, Ghent, Belgium
e-mail: mathias.vancompernolle@imec.be

© The Author(s) 2021   19
G. Concilio et al. (eds.), *The Data Shake*,
PoliMI SpringerBriefs,
https://doi.org/10.1007/978-3-030-63693-7_2

2.1 Introduction and Context

2.1.1 From Smart City to Data City

For over a decade, city governments have been exploring what it can mean to be a "smart city". Considering the centrality of the contemporary urban concept occurring in various discourses on urbanism - from media studies, urban studies, geography, to architecture, and elsewhere - the ongoing role and application of associated ICT within future urbanisation seems inevitable. Turning the promises of the Smart City into practice, however, remains a challenge for cities today. Most agree that technology has some role to play in supporting or implementing policy, but how that role should be filled remains unclear and is often the result of trial and error. It is clear that Smart cities are partly digital, becoming places where information technology is combined with infrastructure, architecture, everyday objects, and even our bodies to address social, economic, and environmental problems (Townsend 2013). The rationale being that, to create a 'better' city, it should be turned into an 'intelligent machine' to both understand and manage complexities of urban life. Connectivity is thus a core feature, as are huge amounts of data collected, generated, and analysed.

The Smart City concept has also been criticized, inter alia for its self-congratulatory tendency, the commercial interests at play, as well as its push of ICT and the potential consequences towards reinforcing a digital divide (Graham 2002; Hollands 2008). Handing over too much control over the public domain to private companies raises concerns regarding democracy and the commodification of the public space (Greenfield 2013; Peck and Tickell 2002; Townsend 2013). Both are a far cry from what would be labelled as "smart". The concept remains fuzzy, meaning different things to different people, from concerns about freedom/privacy to enthusiasm about efficiency, sustainability, economic growth, participation and generally a better world through technology (cf. Komninos and Mora 2018; Mattern 2017).

At the same time, city governments are exploring how the concept can actually contribute to their daily practices and which role technology can play in providing better or "smarter" services to citizens. Even the staunchest critics of the Smart City concept agree that data increasingly has a role to play in policy making (Hollands 2008). Some scholars, therefore, speak of Data Cities rather than Smart Cities (Powell 2014). While this of course has always been the case to greater or lesser extent, the sheer amount of data that is becoming available today, as well as the combination of data from different sources and domains, can provide new types of tools and insights to policy makers. This can be data that comes from Internet of Things solutions (e.g. sensors in public parking garages), structured information in internal reporting systems, detailed data on the public domain (e.g. from satellite imaging) and so on.

In order to fully unlock the potential of this data however, it needs to be more easily available and accessible than today. This is where open data comes in. The idea is that governments currently own (but do not use) a wealth of information related to divergent aspects of life in the city, but that this data is neither publicly available, nor easily interpretable. This has sparked a movement to encourage the opening of datasets in a structured and machine-readable way, under the "open data" moniker, which has gained significant traction across local and national governments. The Open Knowledge Foundation is one of the strong proponents of open data and has come up with what has become the generally accepted definition of open data: *"Open means anyone can freely access, use, modify and share for any purpose (subject, at most, to requirements that preserve provenance and openness)"* (OKFN 2015). This means that open data can be used for any goal at no cost, with the only (potential) exceptions being that reusers mention the source of the data or do not in any way prevent the data from being shared further on.

The idea here is clear: public organizations open up all kinds of data related to their operations, with the goal of having external developers create new services and applications ("apps") based on this data. In principle, this can mean a cost reduction for the public organizations that open data, as they do not need to build and maintain their own services and apps, an activity that is generally accepted as being highly cost intensive (Walravens 2015).

In practice however, a number of challenges remain and "merely" opening up data has not always proven equally successful (see e.g. Peled 2011; Lee et al. 2014). Opening up data already entails significant challenges to governments and public organizations before any data "leaves" the organization (e.g. setting up internal processes to safeguard internal data hygiene and quality control or implementing new or updating existing database systems). Relevant data can also be distributed over different government organizations or levels of governance, and some data applicable to the public may be under the control of private players that are less inclined to open it. After data are made available, the role of government is not necessarily played out. Ensuring that data is actually reused, and relevant applications are built, should also be considered a concern for these public organizations and open data policy makers.

In order to tackle some of these challenges, the Smart Flanders program[1] was initiated by the Flemish Government (Belgium) in early 2017. Smart Flanders is coordinated by IMEC, the largest non-profit technology research institute in Belgium, by an interdisciplinary team of researchers from communication sciences, organizational science, and computer science. The goal of the 3-year program is to support the thirteen so-called center cities in Flanders (by and large the biggest cities) and a representation of the Flemish Community in the Brussels Region (referred to as the 13 + 1), with defining and implementing a common open data policy. The program is followed up by a steering group consisting of representatives of the cities, the cabinets of the Flemish ministers for Urban Policy and for Innovation, the Flemish agencies responsible for Interior Policy and Information, the Knowledge Centre Flemish Cities, the Organization of Flemish Cities and Towns, and IMEC.

To achieve the goal of defining and implementing a joint open data policy, these cities needed to find common ground and collaborate in ways and on themes that were quite new to them. This paper will present some of the most significant challenges

---

[1] https://smart.flanders.be (Dutch only at the time of writing).

at play when it comes to open data in a city context today. It will summarize these points of attention in an Open Data Checklist that cities may reuse to assess their "open data readiness".

2.1.2   Exploring the Cities' Points of View

In order to establish a state of the art around the topic of data/smart cities, a thorough, written, open questions survey was conducted with the cities. This survey asked the participating cities how they looked at the Smart City concept, whether and how they currently organize around it, how they spend resources on Smart City projects and how they think about technology and data. The survey also aimed to document whether any smart city policies were already in place and what these may entail.

This initial written survey was then complemented by a round of in-depth expert interviews with representatives of the 13 + 1 cities. These semi-structured interviews allowed us more insight into the motivations, concerns and challenges raised by trying to establish a smart city strategy. Fourteen interview sessions were held between April and October of 2017, with multiple representatives of the cities present. The profiles that participated in the interviews range from politicians, civil servants responsible for data management, ICT, geographical information, local economy, mobility and so on. Representatives from the following cities were interviewed: Aalst; Antwerp; Bruges; Genk; Ghent; Hasselt; Kortrijk; Leuven; Mechelen; Ostend; Roeselare; Sint-Niklaas; Turnhout and the Flemish Community Commission in Brussels. The interviews lasted between two and four hours and were transcribed for analysis. The data gathered in 2017 (Van Compernolle et al. 2018) is currently being updated during a new round of interviews taking place in Summer 2019. Where possible, we will complement the analysis with this new material. Later publications will focus on these new results and the evolutions we can derive from them over a two-year period.

Based on the insights coming from both this quantitative and qualitative data, a number of critical aspects were identified that cities can actively work on, with the goal of making a smart city and open data strategy more concrete. It became clear that many general challenges remain when it comes to implementing sound open data policies. These challenges came to the foreground during the Smart Flanders steering group meetings and were shared via the website[2] to generate wider debate (in Dutch). The following section will present and discuss these challenges.

---

[2]https://smart.flanders.be.

2.2   Challenges and Questions Related to (Open) Data Policies

The data shake is affecting the policy making domain by giving rise to different challenges in the wider landscape of data policy. Key ones are:

- Data "hygiene": In some cases, digitization still is a significant challenge, but how can we generate awareness to the level of key individual public servants that work with data? How do we change working with data into an operational process that leads to good open data?
- IoT and open data: In the hype surrounding the Smart City concept a lot is made of the data generated by sensors and other IoT devices, but how do we publish data from these sensors in a proper way, dealing with the real-time aspect, the sheer volume of the data, archiving of data and so on?
- Centralization vs decentralization: As a principle, open data lends itself quite well to decentralized publishing and the technical solutions are available, but how do we turn these into processes that work? This requires agreement on the roles of different levels of government.
- Government and the market: where does the role of government end? When do private actors come into play? This is particularly relevant in the field of open data as well.

2.2.1   Data Hygiene in the Organization

The first challenges for most of the cities that were interviewed still relate to the digitization of internal processes and services towards citizens. This also entails having processes and procedures in place when it comes to working with data in the organization. It may seem counterintuitive but open data can actually offer significant short-term efficiency gains in this regard. By reusing data from other organizations or departments within the city, public workers can avoid wasting time looking for the most recent or complete information. This does however require that everyone in the organization that needs to work with data is aware of the importance of doing this in a structured, traceable, and repeatable way. That also means a data management plan at the level of the whole organization becomes an important tool to manage these processes. Very often, this is not or only partially present in the interviewed cities. It is however recognized as being of key importance and is under development in almost all cases. Keeping data hygiene within the organization under control and at a high level is a first long-term challenge and requirement to implement a sustainable open data policy.

Cities also recognize that interoperability will increasingly be of great importance in this context. Making clear agreements on the ownership, use and publishing of data will only grow in importance, but it requires an investment on the part of the organization to ensure sufficient technical expertise and to make the right decisions in this complex area. Interoperability and the concept that data and applications can be seen as separate from each other should prevent data becoming "locked up" in applications provided by third-party vendors. Avoiding so-called vendor lock-in means that the relationship between a local government and its service suppliers can evolve from a typical client-supplier relationship into a partnership in which data are easier to move from one system to another when this is needed or desirable.

2.2.2   IoT and Open Data

The Internet of Things (IoT) is often mentioned in one breath with Smart City services and can mean an extra complicating factor when viewed from the perspective of the data these systems generate. The concept links to the idea that we can understand reality better by measuring as much as possible and by equipping the public space with all kinds of sensors that collect different types of data, policy can be informed by more evidence than ever before. Policy could be tailored to what is observed in the public space, even in real-time.

However, the idea of data-driven policy making comes with a number of complexities on different levels. Divergent actors need to collaborate in new ways and in new fields. One real life example from Flanders is using ANPR cameras to enforce a low emission zone in a city in which certain types of polluting vehicles are not allowed or need to pay a fine when they enter the zone. The sensors in this case are the smart cameras that can detect license plates and determine whether a car can enter the low emission zone or if a fine needs to be sent. To enable this, an elaborate collaboration between different actors needed to be realized, as data needs to be shared between different government organizations, police databases, companies deploying the infrastructure (the cameras in this case) and related software platforms, citizens who need to be informed about which types of cars can enter in the zone during which period and so on.

Next to the often-complex forms of collaboration or partnership between diverse actors, processing all the data generated by IoT solutions is another significant challenge. Clearly, when more sensors are deployed in the city, the amount of data these systems generate increases dramatically. All this data needs to be processed, a task often given to the third-party vendor supplying the solution, but what remains often unclear today is if and how the collected data should be archived. Historical analyses can yield very interesting insights to inform public policy or even allow for predictive analytics, but how long should these large datasets be stored? After which time period should data be erased, especially if personal information is included? Who is responsible for storing and providing access to the data? Who pays for these services? It is important to consider these questions when procuring IoT solutions from third parties and including these arrangements in contracts and agreements. Very often, this is not the case today.

Finally, and to the core of this contribution, a significant challenge related to IoT data is how to publish this data for reuse in a sustainable and cost-effective way. In the spirit of open data, providing potential reusers with real-time information coming from IoT solutions has the potential to generate all kinds of innovative services and applications (e.g. in the domains of mobility, air quality crowdedness and so on). This means however, that infrastructure needs to be made available to allow for a swift processing, publishing, and archiving of said data. Some solutions are available today, but they are often tied to a single vendor or solution. Furthermore, with the speed at which more IoT data is becoming available, this challenge will quickly become more prevalent and need to be addressed sooner rather than later.

2.2.3 Centralization vs Decentralization

Another pertinent challenge or question in the field of open data relates to the way data are published and which actor takes up which role. The question should be framed in a broader debate on centralizing data versus decentralizing them. What remains crucial is that data are easy to find and use for potential reusers. The success of any open data policy will depend on this. Hence, it is important that a local government communicates about the data it makes available, but also that the data can be easily found by anyone looking for it (e.g. also from abroad). When data is published in a decentralized way, for example on the website of the municipality, it is important to describe the data according to standardized principles. By applying standards (like DCAT for example) to describe data, information about that data can automatically be picked up by regional, national and international open data portals, making them easily retrievable by anyone looking to reuse them (including commercial data portals such as Google Dataset Search for example).

Publishing data in a completely decentralized way is technically possible but entails a number of organizational challenges. Clear agreements need to be made about the standards used, the ways in which they are applied and the processes that need to be put in place to ensure data is published in the proper way, for example on a municipality's website. This requires a significant investment by local governments and since open data is rarely a priority, this remains a challenge. Additionally, the resources and skills required are not always present, particularly in smaller organizations. For them, a more centralized approach will prove far more sustainable.

The question then becomes who should take up the role of supporting smaller local governments with this challenge. In Belgium, because of its complex and federated structure, the regional Flemish government, provincial government, or intercommunal organizations could take up this role. Larger cities could take up some of the investment to support the smaller municipalities in their region. And new forms of collaboration between local governments are also coming to the foreground in different regions (e.g. around Brussels). Today, none of these actors are clearly positioned to take up such a role, but it is becoming increasingly clear (and urgent) that more collaboration in this area is needed to set up more sustainable data (sharing) policies.

The first question related to the core competences of government is then; who does what and who has a clear mandate to enforce certain policies if necessary? Today, this situation is fragmented and unclear in Flanders and by extent, Belgium. A broad governance of the Flemish public data landscape should be developed and formalized as soon as possible in order to avoid further fragmentation and an inefficient use of public resources.

2.2.4 Government and the Market

Next to the question of which level of government should take up which role, a second important question related to the core competences of government can be identified: which tasks should be for government and which should be taken up by private players? This is a political decision and choice for the most part and hence will evolve depending on dominant views at the time. As such, it is something of a moving target. This however does not mean this question should not be in the back of the minds of policy makers, as a choice for "more" or "less" government can have consequences for the quality of service provision to citizens.

A key challenge in this area of balancing public and private interest in the context of open data relates to stimulating reuse of open data: should it be a task of the government to ensure that data are actually reused? Most cities agree the local government has a role to play here, by (1) serving as the authentic source for published opened up data (2) ensuring data can be easily found and the threshold for reuse is kept as low as possible and (3) that local government engages in a dialogue with potential reusers so that the data that are published are relevant and of value for reuse. Since data are also made available for commercial reuse, it is not possible to exclude companies from this dialogue. A challenge then becomes how to avoid giving any company a competitive advantage (e.g. by giving them insight into available data or a roadmap for publishing certain datasets). Transparency on both the process and result of a dialogue are crucial here.

Another challenge is the relationship between government and third-party vendors: what are the options as a public organization in enforcing certain behaviour from its suppliers? A number of basic demands can be included in the contracts between the two, are e.g. penalty clauses also foreseen? What is the recourse when the systems of two vendors turn out not to be compatible even though this was ensured during the contracting phase and both suppliers point to each other? Often, local governments do not have the resources to engage in complicated lawsuits. There is no simple answer to these challenges, but the dialogue and transparent approach referred to in the previous paragraph can be part of the answer. Additionally, traditional procurement could be abandoned in some cases where innovative procurement allows for more flexibility on the part of the procuring organization.

A public organization is expected to serve the public interest. When working with and on data, this role becomes even more important, but also far more complex. More than ever, local governments should inform themselves on good practices in this field and clearly position themselves towards third-party vendors that promise the single solution to all of their challenges. By starting from a stronger base of information as well as some shared principles, local governments can evolve away from a traditional client-supplier relationship towards a partnership with market players. When it comes to open data, the role of government here is to strive for a maximal and broad reuse of data, through a transparent process and dialogue.

2.2.5 Open Data Checklist

The survey and interviews with the 13 cities have led to a number of insights related to publishing open data, some of which were outlined in the previous section. To make these insights accessible for reuse by other (local) governments, they are presented as a checklist in what follows. Government organizations that are exploring open data initiatives can use this checklist to ensure to cover some of the most significant challenges related to publishing open data in a sustainable way. The checklist consists of 6 main categories:

- Problem (re)definition
- Capacity and resources
- Organizational culture
- Governance
- Partnerships
- Risks

In Table 2.1, we will very briefly list points of attention in each of these categories.

2.3 Data and Procurement

As hinted at a few times throughout this text, a key tool (local) governments have in this complex context is procurement and the relationship with technology suppliers. During the Smart Flanders programme, it was found that too few, unclear or very different provisions concerning data are included in contracts and agreements with suppliers. In view of the increasing importance of data in the urban context, however, it is extremely important for local authorities to pay attention to agreements concerning data that may be published as open data when awarding public contracts and concessions and when renegotiating existing agreements.

In order to meet this need, a document with model clauses has been drawn up in the context of Smart Flanders. Local authorities can use this when renegotiating existing concessions and public contracts or defining data sharing provisions for new public contracts or concessions with contractors and other third parties. These model clauses are based on the principles of the Open Data Charter, which was also drawn

Table 2.1 Open data checklist

Problem (re)definition	
Frame context and cause	Do not just open data to open data but start from a clear and concrete policy challenge
Define problem and goals	Make the policy goal more concrete by establishing measurable kpis. Open data will never completely solve a problem but can be instrumental in speeding the process along
Do "reuser research"	Understand the needs and pains of potential reusers by engaging in a transparent dialogue
Redefine the problem	Evaluate the initially identified problem and do not hesitate to rescope or redefine it if necessary
Create an overview of the data	Understand which data are available within the public organization and who is responsible for them
Capacity and resources	
Build data infrastructure	Publishing data means the basic data infrastructure needs to function well first. For smaller municipalities this cost can potentially be shared through intergovernmental collaboration
Develop expertise	Working with (open) data requires skills that are today not always present within public administrations. Training and knowledge building in this area is important
Provide sufficient resources	Open data requires an initial investment and a translation into processes within the organization. This requires sufficient means and personnel
Organizational data culture	
Apply shared principles	Whenever possible strive for using shared frameworks so that all partners understand terminology in the same way
Stimulate "believers"	Identify public workers in the administration that see the potential of open data and actively involve them in implementing a policy
Be open for feedback	Reusers of your data will provide you with feedback on data quality, availability and so on. The organization needs to be prepared to tackle constructive feedback
Governance	
Guard standards and data quality	A good internal data hygiene requires the use of standards to allow for easier and automated sharing, linking and exchanging of data
Set roles and responsibility	Clearly defining who does what within and outside of the public organization is key in ensuring efficient use of resources. This is perhaps the most important challenge facing local governments today

(continued)

Table 2.1 (continued)

Problem (re)definition	
Strive towards an agile and flexible organization	Working with data and technology requires flexible processes to allow for corrections when needed
Develop structured evaluation	Foresee quantitative and/or qualitative kpis to evaluate both process and outcome. This means including a baseline measurement as well
Partnerships	
Approach data owners	Explore new partnerships with owners or relevant data to support policy challenges
Involve domain experts	Include the domain expertise present in the public organization to ensure data is described and applied in correct ways
Involve organizations with similar goals	Use the knowledge and expertise of like-minded organizations, whether they be other local governments, departments within other levels of government, civil society, companies, research centers and so on
Procurement	When procuring new solutions or renegotiating contracts with third-party vendors, include clauses related to data ownership, processing, storage and open data
Risks	
Privacy	Develop privacy-by-design solutions and applications and include privacy impact assessments when publishing data. Open data per definition does not include personal data, however scenarios could be envisaged where the combination of open data results in the identification of individuals. An a priori privacy impact assessment can identify this
Security and data management	As local governments start processing more data, security becomes increasingly important as well. A data management plan can support this but may require external capacity and support
Digital exclusion	Open data initiatives should never lead to an exclusion of those who do not have the skills or access to public services
Data quality and policy decisions	Evidence-based policy can only be as good as the data that support it. Data quality and verification are thus of high importance, also when opening up. A guiding principle here can be that if data are considered of sufficient quality to be used internally for policy development, they should be of sufficient quality to open up

(continued)

Table 2.1 (continued)

Problem (re)definition	
"Open washing"	This risk refers to a situation in which public organizations claim to open up, but only do so to comply with regulations. This is not a sustainable situation and waste of resources. Starting from a concrete case or project can avoid this

up during the Smart Flanders programme. The Open Data Charter contains twenty general principles that together form the ambition of the 13 centre cities of Flanders. The Charter was also adopted by the government of Flanders and is available online.

The purpose of the model clauses is, among other things, to give the city organisation direct access to data and to regulate the responsibilities with regard to the publication of these data for re-use. In addition, a more uniform approach to data in tendering is provided. The main target audience of the document is local contracting authorities, but it can also be used by other contracting public authorities. The document has been conceived as a sort of guide, first briefly explaining what (open) data and linked open data are, and why it is important to consider them during procurement. It also follows the structure of a typical specification document, referring to selection criteria, award criteria and technical criteria. This distinction is of course critical when drawing up procurement specifications and authorities can decide to what extent they want to use the model clauses in one of these categories, depending on the solution that is being procured.

2.3.1 Examples of Model Clauses

By way of example and with the goal of inspiring others to take up similar initiatives in their localities, some of the model clauses are included below. It should be noted however that these are merely translated from Dutch, in accordance with existing legislation applicable within Flanders, and should be checked for conformity to local applicable law.

The contracting authority starts with the delineation of open data:

"To make public and private information services possible, (static and dynamic) open data are essential, and this in all areas of policy making. The contracting authority therefore endorses the principle that all datasets, data and content that anyone is free to use, adapt and share for any purpose are referred to as open data, with the exception of those data and datasets of which the confidentiality is protected by law or may logically be expected, such as personal data, data compromising public order and security (hereinafter "Open Data")".

Subsequently, the contracting authority must indicate which data must be collected and consequently possibly published as open data:

"The contracting authority entrusts the contractor with the collection of the following data (hereinafter the "Collected Data"):

- [*to be completed by the contracting authority*];
- …"

The "Collected Data" that the contracting authority expects to be collected by the contractor (and of which the contracting authority becomes the owner) should be described and listed in as much detail as possible. After all, the contracting authority should only have the data that is relevant to it, be collected by the contractor.

In order to ensure that the data are eligible for re-use, it is important that clear agreements are made about the ownership of the data:

> "*The procuring authority owns the Collected Data. The Contracting Authority has the right to copy, distribute, present, reproduce, publish and reuse the Collected Data. The Contracting Authority must have immediate access to and be able to make full use of the raw Data collected by the Contractor, both during and after the term of the Contract. This also applies to historical data. The Contractor may still use the Collected Data itself for the purposes for which it deems it necessary.*"

Finally, for the purpose of this paper, data quality can be an important factor as well. The model clauses give the contracting authority the possibility to impose quality requirements on the contractors with regard to the Collected Data.

"*At the request of the contracting authority, the contractor shall make available to the contracting authority a provisional version of the datasets, as well as the URLs referring to the opened datasets;*

- *During the performance of the contract, the contracting authority may have the 'Collected Data' verified by an external party designated by the contracting authority;*
- *If the contracting authority makes use of the review described in the previous paragraph, the contractor has the opportunity to follow up on any comments;*
- *The final result will be inspected after the Contractor indicates to the Contracting Authority that the final result has been achieved.*
- *During this inspection, the leading official or his authorised representative checks the quality of the Collected Data by means of a general, technical and content-related quality check. During this inspection, the conformity of the format and the semantic aspects of the standard as well as the conformity with the technical specifications, such as the completeness, correctness, positional accuracy and timeliness of the Collected Data are checked.*
- *The Contractor is obliged to comply with the remarks made to him by the leading official or his authorised representative*".

If the result of the inspection shows that there are defects in the way in which the Collected Data were published, the contracting authority can opt to have these defects rectified by the Contractor, if it considers this to be appropriate. If the contractor fails to take remedial measures, the contracting authority may take an ex officio measure at the contractor's expense and risk. In addition, the contracting authority may include special penalties in the contract documents, which may be imposed if the contractor fails to take remedial measures.

Again, these examples only serve as an inspiration and should be adapted to the local context. The guide continues with a set of technical model clauses on how access to the data should be organised, how data can be published in a decentralised way, how sustainability of the published data should be organised, which metadata and other standards can be used and so on. The full document (in Dutch) is available with the authors by simple request or on smart.flanders.be.

2.4   Discussion and Conclusion

In its most basic operationalization, the idea of opening up government data where possible, holds a great deal of potential. It can give citizens more insight into how and why certain decisions are made at the political level. It can also stimulate innovation, with new services, apps, efficiency gains, jobs, and economic activity as a result. Lastly, it can lead to more and better interaction between citizens and governments and so on.

Yet, it is questionable whether it actually does, or what the conditions should be for this to take place. This is made explicit in the political premise of the Smart City concept, in how politicians frame their view on Open Data: the concept is quite popular across the political spectrum, as it can be employed in very different rhetoric; as an argument for a smaller government (not building services and applications, but making sure data are available so that others can do so) or one for more government effort (e.g. in relation to transparency, engagement with citizens, active participation, development of data-related software solutions, standardization activities and so on).

Whatever viewpoint taken, a "Smart City" should include, at the very least, access to data. However, as reuse of open data does not 'just happen' and requires interaction, stimulation or incentives in some cases, the question becomes at which point the role of public officials is played out in this realm (Walravens et al. 2018). It is clear that the government body providing open data has a role to play, but to which extent? In what forms should it make data easily available, but also understandable or interpretable for citizens? For which types of data or in which domains? How can open data be privacy-compliant? Herein lies the potential for a—perhaps counterintuitive—democratic deficit of open data: even if data are available in a Smart City context, it does not mean they are "usable, useful or used" (Open Knowledge International 2019).

One part of the answer seems to lie in avoiding a purely top-down or bottom-up approach (Shepard and Simeti 2013), but rather aiming to bring together the relevant parties from the quadruple helix (government, companies, research and citizens) as mentioned above. Engaging the quadruple helix, and particularly citizens - via truly participatory and inclusive means, in complex urban challenges with technical components like (open) data - remains a massive challenge. Such an approach requires sufficient time and means to facilitate discussion, properly defining urban challenges, getting the roles of all involved stakeholders clear and setting up a step-by-step approach to act. Only through such an approach can a more sustainable open data policy be developed, that further enables a Smart City.

Opening data remains something of a chicken-and-egg problem: sufficient investment is needed on the side of the government in order to publish significant amounts or relevant data, but reusers will only generate innovative applications and services once enough data are available.

The research presented in this paper shows that cities certainly see the potential value of open data, but a number of challenges remain. In order to develop sustainable open data policies, a number of conditions have to be met. These have been summarized as points of attention presented in an Open Data Checklist. Additionally, procurement is a key process in government which unfortunately does not always sufficiently take provisions on data into account. This contribution illustrated how using the same model clauses related to data, open data and linked open data can create benefits for both contracting government organisations, as well as technology suppliers. Taking factors related to problematization, organizational culture, governance, partnerships and a number of risks into account, as well as optimising procurement strategies can help local governments make more informed decisions when designing or developing an open data policy for their constituency.

References

Graham S (2002) Bridging urban digital divides: urban polarisation and information and communication technologies. Urban Stud 39(1):33–56

Greenfield A (2013) The city is here for you to use. Wired, 5 February

Lee MJ, Almirall, E, Wareham JD (2014) Open data & civic apps: 1st generation failures–2nd generation improvements. ESADE Business School Research Paper, (256)

Hollands R (2008) Will the real smart city please stand up? City 12(3):303–320

Komninos N, Mora L (2018) Exploring the big picture of smart city research. Sci Reg 15–38 (ISSN 1720-3929)

Mattern S (2017) Code + Clay…+ Data + Dirt: Five thousand years of urban media. University of Minnesota Press, Minneapolis

Open Knowledge International (2015) Open definition. Retrieved from http://opendefinition.org

Open Knowledge International (2019) What is open? Retrieved from https://okfn.org/opendata/

Peck J, Tickell A (2002) Neoliberalizing space. Antipod 34(3):380–404. Doi:10.1111/1467-8330.0024

Peled A (2011) When transparency and collaboration collide: the USA open data program. J Am Soc Inf Sci Technol 62(11):2085–2094

Powell A (2014) 'Datafication', transparency, and good governance of the data city. Digital enlightenment yearbook 2014: social networks and social machines, surveillance and empowerment, pp 215–224. https://doi.org/10.3233/978-1-61499-450-3-215

Shepard M, Simeti A (2013) What's so smart about the smart citizen. Smart citizens, 4

Townsend A (2013) Smart cities. Norton & Company, New York

Van Compernolle M, Waeben, J, Walravens N (2018) Eindrapport Smart Portrait. Public Report for Kenniscentrum Vlaamse Steden and Agentschap Binnenlands Bestuur. Available at http://www.kenniscentrumvlaamsesteden.be/overhetkenniscentrum/Documents/Eindrapport%20Smart%20Portrait_PUBLIEK%20-%20definitief.pdf

Walravens N (2015) Mobile city applications for Brussels citizens: smart city trends, challenges and a reality check. Telemat Inform 32(2):282–299

Walravens N, Van Compernolle M, Colpaert P, Dumarey N (2018) Open data: opportuniteiten en Uitdagingen voor Lokale Besturen. Politeia

Nils Walravens graduated cum laude as Master in Communication Sciences at the Free University of Brussels in July 2007 and obtained his PhD in the same field in October 2016. As a senior researcher at imec-SMIT, his work focuses on the interplay between the private and the public sector in the areas of innovation and public service provision, researching topics such as business models, value networks and governance models in the context of smart cities and open government. Nils coordinated the Smart Flanders programme (2017–2019), funded by the Flemish Government, supporting the 13 centre cities of Flanders on strategic and technical aspects related to open data. He currently works on several projects related to open data, digital twins and open innovation in the public sector.

Pieter Ballon is Senior Researcher and Professor of Media Studies at Vrije Universiteit Brussel. He is a specialist in business modelling, open innovation and the mobile telecommunications industry. Formerly, he was senior consultant and team leader at TNO. In 2006–2007, he was the coordinator of the cross issue on business models of the Wireless World Initiative (WWI), that united five Integrated Projects in the EU 6th Framework Programme. Currently, he is the international Secretary of the European Network of Living Labs. He holds a PhD in Communication Sciences and a MA in Modern History.

Mathias Van Compernolle is a researcher at MICT, an imec research group on Media, Innovation and Communication Technologies in the Communication Science department at Ghent University. In his research he focuses on the electronic and digital government, open government data, smart cities and (network) governance within data projects.

Koen Borghys is a researcher at imec-SMIT focusing on (open) data, data driven governance and smart cities. Koen graduated in 2012 as Master of Laws (LL.M) and in 2014 cum laude as Master of Science in General Economics (2014). After his studies, for several years he worked as a lawyer advising young technology companies then becoming a researcher at imec-SMIT. He currently works on several projects related to (open) data, data driven governance and smart cities, including the VLAIO projects 'Smart Retail Dashboard' and the VLAIO COOCK project 'Open City'

Chapter 3
Towards a Public Sector Data Culture: Data as an Individual and Communal Resource in Progressing Democracy

Petter Falk

Abstract An increased use of data has swept through many policy areas and shaped procedural and substantive policy instruments. Hence, citizens and governments, as both producers and consumers of data, become intertwined in even more complex ways. But the inherent logic of data-driven services and systems sometimes challenges the prerequisites and ideals of liberal democracy. Though a democratically sound data-practice and data-culture is crucial for ensuring a democratic usage of citizens data, discourse tends to overlook these aspects. Drawing on insights from the project Democracy Data, this chapter explores the opportunities and obstacles for establishing democratically oriented public sector data cultures.

Keywords Data-culture · Public sector · Democracy · Value-creation · Design · Conceptualization

3.1 The Balance of a Data-Driven Democracy

The balance between individual and collective needs and interests being acknowledged and addressed is a liminal factor in the practices of any liberal democracy (Dahl and Shapiro 2015). Be it protection, taxes, rights or liberties—government and politics is calibrating this equilibrium in order to create value and allocate resources. Digital data, as a potential recourse, is one that both citizens and governments possess. And as the public sector is entering into an era of algorithmic governance—where algorithms, automation and data-driven services constitute a cornerstone in decision making (Doneda and Almeida 2016), that balance needs a new articulation (Keller et al. 2017).

In order to create value for both *the citizens* and *the communal* there needs to be some form of integration between the resources of the individual and the resources of the government (Vargo et al. 2017). Digital data as a resource on an individual level can help the individual inform personal decisions, or help a government professional,

P. Falk (✉)
Research Institutes of Sweden, Karlstad University, Karlstad, Sweden
e-mail: petter.falk@ri.se

© The Author(s) 2021
G. Concilio et al. (eds.), *The Data Shake*,
PoliMI SpringerBriefs,
https://doi.org/10.1007/978-3-030-63693-7_3

like a doctor or social worker, to provide sufficient care. At the same time, that very same singular information-point, when aggregated, can be a collective good. It helps inform decision makers as statistics or real-time data. But an extensive use of individual and aggregated data also carries many risks. On an individual level these range from the infringement of personal integrity to direct or indirect discrimination. On an aggregated level; statistical fallacies or false reliance from not accounting for data quality or causality (Loukissas 2019). So, in sharing data as a resource, as with any democratic value-creating process, balance is needed. Though unlikely such perfect balance will be reached as a perpetual state, or for that matter come to fruition, what the conceptualization and reiteration of democracy has taught us is that striving for democracy as an ideal is as close as one can get (Dahl and Shapiro 2015). And as the public sector is becoming more data-driven some argue that governments need to take a technology-assessment approach to digital tools and concepts in the democratic discourse (Poullet 2009; Nemitz 2018). How does technology shape, contribute to- or disrupt the constitution of democracy? In this regard, articulating an understanding of data on both the citizen's and the government's terms is crucial not only to ensure a representative usage of the public's data, but also to foster a sustainable approach to development and innovation suitable for a public sector logic and a democratic agenda.

It has been said that the strangeness of data is it's strength (Loukissas 2019). Unlike monetary value, where one tax-euro can only be spent on one thing, data can be used and reused in perpetuity. And this strangeness, data's ability to be many things in many settings, and be a resource in more than one way, is becoming an imminent part of our ongoing discussion on democratic ideals. This chapter, in the light of this narrative, illustrates how the data-practice and data-culture of public sectors impacts the balance between individual and communal value-creation. It does this specifically from an administrative point of view, looking at both the infrastructure implementation of data-driven systems and services, concentrating on the data produced directly or indirectly by citizens' usage of welfare service. And, drawing from the research- and innovation project Democracy Data (VINNOVA 2018), it proposes a sequence of interwoven tactics intended for policy makers, public sector managers and data-practitioners in public administration for furthering a democratic tenacity in the practice of government data-culture.

3.2 The Conflicting Logics of Emerging Public Sector Data Cultures

In theory it's kind of simple. When the citizen interacts with the public sector in a digital interface, either directly or indirectly, it leaves data as a digital footprint. After this spark of data-creation, data rests in datasets or databases, transits between human and non-human agents or is put to use (Nelson et al. 2009).

However, the use and understanding of data is not bereft of history, norms and hereditary logics (Bates 2017), a circumstance that is especially significant in a democratic framework. Even though computing as a form of practice for processing information in governments has been around since at least the 1930s (Wynn-Williams 1931), the advent of interconnected computer networks lay the groundworks for governmental data practice back in the 1990s (Ho 2002). Nevertheless, it has not been data as an artefact or resource, but the technological implementation and innovation which has been in the limelight of digitalization discourse. In the emerging days of the Internet, it was theorized that technology would enable new visionary forms of digital culture, empowering direct democracy and removing participatory barriers by means of novel technology (Rheingold 1994; Dyson 1997). However, public sector organizations where more inclined to make incremental changes in existing services and operations using these emerging technologies (Norris 2003). Drawing from commercial rationalities on usage and implementation and merging them with governmental undertakings created concepts like e-government, (Layne and Lee 2001) government e-services (West 2004) or e-democracy (Chutimaskul and Funilkul 2004). Concepts like these lay the initial groundworks for a shared understanding of the digital relationship between the individual citizens and the government (Fang 2002). And within this jargon, the citizens wellbeing, access to- and engagement with their government is very much reliant on the capacity of the user to interact accurately with the system at hand (Jaeger and Bertot 2010), meaning that value was created *in action*, relying on the agency of the citizen. However, if one looks past the front-end interfaces and technical infrastructure and focuses primarily on data, the notion of agency is less articulated. Of course, governments can do a lot of things with data without the citizens' presents or immediate action. But how does this practice fit into the larger purpose of the public sector?

Given this question, and as more professions, practices, and decision making processes revolved around digital data, and as Big Data moved from viable concept to reality, the term *data-culture* emerged as a ductile concept (Bates 2017). Behaviours, norms, institutions and knowledge dictating data-practice in a given context are factors that in turn order the prerequisites and ambitions under which data is accumulated, processed and decimated (Kitchin 2015). As any given culture, the dialectic nature between cultivation and organic growth shapes its assemblages, rationalities and realities. The socio-material context and surrounding shape the subject, and it also shapes the understanding of the subject. But directed efforts allow for a group or network to foster or shape their practices through both social and technical factors. As such, local data cultures are created, sustained and transformed by existing in a given environment and at the same time interacting with adjacent digital and social systems. This is what constitutes them (Bates 2017). There is of course no universal public sector data culture. As with all institutions, politics and government included, norms and beliefs that dictate social codes are multi-layered arrangements where different local cultures share similarities but also display differences (Hall and Taylor 1996). As such, data-cultures have emerged within the modern-day government, both on national and local levels. These cultures are constituted by the socio-material conditions and practices of digital tools and services that generated and accumulated user-

and meta-data through a multitude of digital interactions throughout the public sectors digital dispositif. And within the context of any government, such as the local ones, a multitude of data cultures, with unique expressions, norms and practices, can be observed (Bates 2017). There are however, arguably, traits and logics that—from the vantage point of liberal democratic welfare states—are recurrent in most cultures. One such central theme is goal of value-creation for the citizens, a topic that in a digital framing has been covered extensively (Grimsley and Meehan 2007; Ebbers 2016; Nielsen and Persson 2017; Lindgren et al. 2019). In order to create value for its citizens, the democratic government enacts certain arrangements and ideals, such as digital participation, transparency and accountability, improving e-services and using technology to reduce public spending (Jaeger 2005). Another shared trait, if not by all then by the most progressive public sector organizations, is the ambition to hold the competence to understand the data-driven process, or what's commonly called data-literacy (Markham 2020).

What these shared traits of public sector data-cultures illustrate is a need to conceptualize how democratic value is created in a data-driven public sector. But in this conceptualization process there is an inherent and probable prospect of conflicting logics influencing perceptions and foci. In general, we want to assume that the public sector has the ambition to create value for its citizens on democratic basis (Dahl and Shapiro 2015). And even though governments might have been able to do this in a direct way relying on the agency of its citizens, there are norms and hereditary logics in data praxis and digital ventures that can be in conflict with the organizations democratic ambition when it comes to data practice. As much of the practice and culture of data is influenced by the commercial backdrop from which the technologies, languages and enactments of digital discourse has emerged, technology in the public sector is destined to be immersed in commercial logics. This becomes paradoxical in a public sector setting, for example if applying a commercial digital logic to a public welfare service. For example, if commercial data-centric services like Facebook, Google and TikTock using non-linear business models, can create commercial value from user-data, a conventional question would be if public services could create social value using similar data-driven positions, practices and logic? But are such assumptions of public value even apt? Services in a commercial logic is steeped in assumptions from commercial actors, framing the service as situated in a functioning market and as a neutral or positive pursuit. But many governmental services are inherently negative, meaning that services such social welfare and health care generally are service deemed for those in need (Morgan and Rao 2003). Parallel with this, contemporary mainstream data-driven technology is, as mentioned, inherently interconnected with a commercial logic. Systems are developed and procured from commercial agents and the discourse and zeitgeist has, in many ways, been shaped by the companies and innovators who develop the systems. This not only affects how things are done, it also shapes how we understand realities. As the practices of data reflect the ontological positions of its practitioners, discourses and socio-material context inherently also affects ethical and conceptual positions (Bates 2017). Thought research acknowledges the importance of data-literacy in public sector value creation in order to uphold certain acknowledged democratic principles (Markham 2020), the question of how to foster a culture that approaches the multiplicity of data as both an individual and communal recourse, honouring and furthering democratic discourse, is far removed from the contemporary practice and setting of data.

3.3 The Project Democracy Data—Lessons on Cultivating Local Data Culture from the Swedish Social Services

The project Democracy Data was initiated in the fall of 2018 to explore the practices, understandings and cultures of data in the social welfare offices of two Swedish cities; Malmö and Karlstad. The project followed and monitored the implementation and iteration of a series of data-driven services and systems throughout a time span of 15 months. Through a mix of para-ethnographic methods, interviews and surveys the project looked at the socio-material context of resources, technology, policy, economy and history constituting local data culture within the municipal social services.

There are a few things to be said about the municipal social welfare administrations in Sweden. In many ways this administrative branch represents the ultimate government safety net in a community. Social work carries a culture of social pathos, stemming from the fact that social work is very much a form of affective practice within a government setting (Penz and Sauer 2019). However, social welfare services in Sweden in general have no map for understanding how, when and where data should and could be used to create social value, beyond the use of mere statistics. Literacy in their data-practice, the project found, is limited, meaning that the complexity of the data-driven systems used on a day-to-day basis (such as electronic records) are often acknowledged, but few have the knowledge or mandate to explore them beyond interface level. Previous research has pointed out that the epistemology in the intersection between the citizens and services, a milieu that very much illustrates social welfare services, often ignores both the context and complexity that manifests in the meeting between the citizens everyday life and the public sector (Madsen et al. 2014). Digitalization of services can add to this complexity, but it can also create a basis for making sense of both individual and public value (Sklyar et al. 2019). As a municipal branch contemporary social work also illustrates the complexity of data-driven service arrangements within policy-driven organizations and networks. Even before the advent of digital technology, municipal social services were part of a vast information-ecology, tying together social security, healthcare, tax agencies, NGO's and the judicial system (Svensson 2019). And historically, local administrative branches have been reliant on the central city management in procurement and development of technical services and substructures. This means that the ownership of implementing, iterating and articulating the use of digital systems has been very much dominated by the city management IT departments, meaning that the practice of data not only was influenced by the commercial logics of the system developers, but also the municipal IT-department as an intermediary.

In the case of both the cities Karlstad and Malmö, the context and the baseline for perceived success in cultivating democratic local data-culture was very much reliant on a data-literate leadership as well as the ability and time to explore data beyond the real and perceived ownership of data-driven systems and processes. Though the concept of data as both an individual and collective value-creator is not articulated, stakeholders in and around these processes voiced a need to situate themselves as experts in their own data. But as legal frameworks are changing and adapting around data ownership, the ability or mandate to explore data is attentively regulated. For example, on a European level, the General Data Protection Regulation in theory is meant to ensure the individual some form of ownership of one's data (EU 2016/679, n.d.). But given that many are unaccustomed to thinking about and conceptualizing data, the chance of malpractice in relations to GDPR discouraged practitioners within the social service fom advancing the current data-practice beyond the acting norms.

And just as democracy relies on the balance between individual and collective needs and interests, data-culture is a balance between organic growth and cultivation. So how to foster a policy logic for the public sector that maximizes data as an individual and communal resource? And how to do it in a way that takes into account socio-material context and a progressive approach to democracy? Drawing on the insights from Democracy Data, tying together the actualities on conflicting logics, the cognizance of data's latent risks and potentials as well as broadening the reasoning on democracy in the digital age, three proposals can be made.

3.3.1   Proposals 1: Promote Holistic Data-Literacy

It has been said that all data models are false, but that some are useful (Box 1976). Understanding what models are useful intrinsically means understanding the origin of data (Loukissas 2019). A holistic data-literacy, taking the alignment from context and accumulation of data to dissemination and decision making into account, hence is crucial for safeguarding individual and cultural wellbeing (Markham 2020).

Data-literacy, as understood by a conventional logic, is generally tied to doing data-work, such as modelling. However, the skillset to draw insight from data other than models and scale, and finding depth in insights, often requires going beyond the numeric realm, circumventing the view from nowhere (Haraway 1988). For example, in a given dataset representing a community, some citizens are bound not to be represented, especially if the data is based on digital interactions between citizens and government. As with any form of active participation, social capital impacts on the actual usage of data driven tools and service, especially in a government context (Naranjo-Zolotov et al. 2019). So, if a government agency is to iterate one of its digital services based only on existing user-data, then the improvement of the iteration will predominantly yield the existing user-base. The implication of such fallacy of data-literacy, in an era of automation and AI, risks disenfranchising already marginalized groups within a community, as they are not represented in the accessible data.

Here, it is important to promote and cultivate a data-literacy where you situate your practice in the socio-material context. Loukissas (2019) frames it as going *from datasets to data-settings*, exploring and examining the space for data accumulation together with data-stakeholders and policy makers. In practical terms, this means doing excursions, in-depth user-research or just meeting and interacting with the material-semiotics agents of a data dispositif. It might seem like a banal enterprise, but it inhabits a crucial aspect of the data-literacy skills; a critical view of causality. In public services, especially welfare services, data is being accumulated through the interaction between citizens and the service provider. You interact with your doctor, nurse or social service officer, where in an ideal circumstance, the both of you are trying to frame your needs consequently. In that data-setting, there are stratums of tacit information—language, power, gender, knowledge, bodies, etc. And the one appliance to translate this setting into digital data is the electronic journal system. A system that translates this interaction into a log (for QA and future decision making regarding you as an individual) and statistics (for knowledge and future decision making). If you as a data stakeholder, be it policy makers or analyst, don't situate yourself in that data-setting as part of your data-literacy ambition, there is a risk of losing the depth of insights needed to make relevant and justifiable decisions and policy.

3.3.2 Proposals 2: Design Your Data-Driven Services as if Democracy Depended on It (Because It Does)

Services matter in a democracy. If citizens don't believe that the services and utilities provided by their government is delivering value, they lose faith in the system (Peters 2010). And design matters in services. If a service design fails to capture the needs of its user, then there is no purpose of the service. And culture matters in design, as social structures arguably are crucial materials of holistic service design (Vink 2019).

Acknowledging that socio-materiality matters in the design process suggests that organizations can adapt a more all-encompassing understanding of the impact of its services (Akama 2015; Kimbell and Blomberg 2017). Citizens data, potentially, can be both an individual and collective good. And most often, the ability to use data right starts at accumulation. In service implementation, we have the rare ability to orchestrate accumulation of data in unique ways. But in order to maximize the use of data you need to design the space where digital data originates in a holistic way, drawing on the socio-material context, as well as contemplating where data as a resource could and should be used in aggregate forms. As the digital data accumulated today might last forever, and planning for perpetuity is no small matter, such designs need to be situated in agreement between both the citizen and the government. If data-driven service design is assuming that value is embedded in tangible outputs or exchanged between actors (Vargo and Lusch 2004), and we want to make it democratic and sustainable, it has to thoroughly acknowledge the complex systems into which the data is being propelled. This sober cognizance might be overwhelming. But instead of paralyzing data-practitioners and stakeholders, this realization can be used to elevate the role of design in both data-literacy and data-culture, and also to involve the citizens and their perspectives and realities in the discourse of one's data-culture.

Drawing from the research of Vink, there are a set of underlying assumptions in service design and development that is inhibiting certain cultures, organizations or networks from adopting a more situated understanding of design (Vink 2019). Working to shift these assumptions through practice and culture helps understand the gravity of designing a data-driven service, but also supposing a more holistic perspective on digital democracy. In practical terms this means adopting design practices and methodology among data practitioners and stakeholders, but also making sure that design of data-driven services and systems happens in dialogue not only between governmental branches, but also between citizens and government.

3.3.3   Proposals 3: Conceptualize Data as Democratic Artifact

Representation is central to the idea of democracy (Dahl and Shapiro 2015). If the structures and culture of a government do not allow for its citizens to be represented, through voting or public dialogue, then governments, in theory, lose its democratic legitimacy (Zittel and Fuchs 2006). So, as the role of data in public life is changing the government's work and as data represents citizens in a multitude of ways, how the government enacts the citizen needs reframing (Jaeger and Bertot 2010).

Conceptualizing future roles of technology always is done balancing between an utopian and dystopian rhetoric (Boyd and Crawford 2012). With the risk of slanting into both these realms it should however be emphasized, based on previous research and the insight from Democracy Data, that a data-driven public sector needs to view citizens data as both a democratic artifact and a conversation between citizens and the government. Data as *enacting* and *representing* the citizens. This notion of conceptualizing data as democratic artifact as well as a conversation between citizens and the government requires an established democratic foundation, including a holistic data-literacy and a design-oriented approach to value-creation.

Data as a democratic artifact means it manifests the citizen through a sort technique of representation. Just as a vote in a general election or a dialogue with elected officials tells us something about the position of the citizen and the community, so does data. If one draws from the actor-network-theory approach of translation, this could mean a process where a set of actors, human and non-human, become proxies for a multitude of other actors, where manifestations can be made based on articulating and linking identities in simplified or fixed forms (LaTour 1999; Law 1999). Data hence could represent one or many actors, and hence be both an individual and collective resource. And by avoiding the customer-oriented logics of commercial ontology, data—on a conceptual level—can be voiced as a source for constituting democracy as well as contributing to both individual and communal value-creation.

Dwelling on such conceptualizations, of course, for many public sector operatives would be an indulgence, given austerity measures and the sometimes harsh realities of contemporary public servants. However, in order to advance a data-culture that helps to articulate new and digitally relevant perspectives on democracy, decoupled from past and commercial logics, a new ontology is needed. Hence, appreciating data as both an individual and communal recourse, is but another step in the balance of advancing liberal democracy.

References

Akama Y (2015) Continuous re-configuring of invisible Social structures. In: Bruni EA, Parolin LL and Schubert C (eds) Designing technology, work, organizations and vice versa, Vernon Press, Malaga, pp 163–183

Bates J (2017) Data Cultures, Power and the City. Data and the City, 189–200. https://doi.org/10.4324/9781315407388-14

Box GEP (1976) Science and statistics. J Am Stat Assoc 71(356):791–799

Boyd D, Crawford K (2012) Critical questions for big data. Inf, Commun & Soc 15(5):662–679. https://doi.org/10.1080/1369118x.2012.678878

Chutimaskul W, Funilkul S (2004) The framework of E-democracy development. Lecture notes in computer science, pp 27–30. https://doi.org/10.1007/978-3-540-30078-6_5

Dahl RA, Shapiro I (2015) On democracy. Yale University Press. https://play.google.com/store/books/details?id=5aYXCgAAQBAJ

Doneda D, Almeida VAF (2016) What is algorithm governance? IEEE Internet Comput 20(4):60–63. https://doi.org/10.1109/mic.2016.79

Dyson E (1997) Release 2.0: a design for living in the digital age. Broadway Books. https://books.google.com/books/about/Release_2_0.html?hl=&id=eFgy1GFW42YC

Ebbers WE, Marloes GMJ, van Deursen AJAM (2016) Impact of the digital divide on E-government: expanding from channel choice to Channel usage. GovMent Inf Q 33(4):685–692. https://doi.org/10.1016/j.giq.2016.08.007

EU 2016/679 (n.d.) Regulation (EU) 2016/679 of the European parliament and of the council of 27 April 2016 on the protection of natural persons with Regard to the processing of personal data and on the free movement of such data, and repealing directive 95/46/EC (General Data Protection Regulation) (Text with EEA Relevance). http://data.europa.eu/eli/reg/2016/679/oj

Fang Z (2002) E-government in digital era: concept, practice, and development. Int J Comput, Internet Manag 10 (January)

Grimsley M, Meehan A (2007) E-government information systems: evaluation-led design for public value and client trust. Eur J Inf Syst 16(2):134–148. https://doi.org/10.1057/palgrave.ejis.3000674

Hall PA, Taylor RCR (1996) Political science and the three new institutionalisms. Polit Stud 44(5):936–957. https://doi.org/10.1111/j.1467-9248.1996.tb00343.x

Haraway D (1988) Situated knowledges: the science question in feminism and the privilege of partial perspective. Fem Stud 14(3):575. https://doi.org/10.2307/3178066

Ho AT (2002) Reinventing local governments and the E-government initiative. Public Adm Rev 62(4):434–444. https://doi.org/10.1111/0033-3352.00197

Jaeger PT (2005) Deliberative democracy and the conceptual foundations of electronic government. GovMent Inf Q 22(4):702–719. https://doi.org/10.1016/j.giq.2006.01.012

Jaeger PT, Bertot JC (2010) Designing, Implementing, and evaluating user-centered and citizen-centered E-government. Citizens and E-Government, pp 1–19. https://doi.org/10.4018/978-1-61520-931-6.ch001

Keller S, Lancaster V, Shipp S (2017) Building capacity for data-driven governance: creating a new foundation for democracy. Stat Public Policy 4(1):1–11. https://doi.org/10.1080/2330443x.2017.1374897

Kimbell L, Blomberg J (2017) The object of service design. Designing for Service. https://doi.org/10.5040/9781474250160.ch-006

Kitchin R (2015) The data revolution: big data, open data, data infrastructures and their consequences. Sage, London

Latour B (1999) On recalling ant. Sociol Rev 47(1_suppl):15–25

Layne K, Lee J (2001) Developing fully functional E-government: a four stage Model. GovMent Inf Q 18(2):122–136. https://doi.org/10.1016/s0740-624x(01)00066-1

Law J (1999) After ANT: Complexity, naming, and topology. In: Law J, Hassard J (eds) Actor network theory and after. Blackwell, Oxford, pp 1–14

Lindgren I, Madsen CØ, Hofmann S, Melin U (2019) Close encounters of the digital kind: a research agenda for the digitalization of public services. GovMent Inf Q 36(3):427–436. https://doi.org/10.1016/j.giq.2019.03.002

Loukissas YA (2019) All data are local. The MIT Press, Cambridge, MA. https://doi.org/10.7551/mitpress/11543.001.0001

Madsen CØ, Bull Berger J, Phythian M (2014) The development in leading E-government articles 2001–2010: definitions, perspectives, scope, research philosophies, Methods and recommendations: an update of Heeks and Bailur. Lecture Notes in Computer Science, pp 17–34. https://doi.org/10.1007/978-3-662-44426-9_2

Markham AN (2020) Taking data literacy to the streets: critical pedagogy in the public sphere. Qual Inq 26(2):227–237. https://doi.org/10.1177/1077800419859024

Morgan I, Rao J (2003) Making routine customer experiences fun: some companies have discovered the competitive advantage of injecting the element of fun into traditionally neutral consumers environments. MIT Sloan Management Review, Gale Academic OneFile 45(1):93

Naranjo-Zolotov M, Oliveira T, Cruz-Jesus F, Martins J, Gonçalves R, Branco F, Xavier N (2019) Examining social capital and individual motivators to explain the adoption of online citizen participation. Futur Gener Comput Syst 92:302–311. https://doi.org/10.1016/j.future.2018.09.044

Nelson DE, Bradford WH, Croyle RT (2009) Making data talk: the science and practice of translating public health research and surveillance findings to policy makers, the public, and the press. Oxford University Press. https://play.google.com/store/books/details?id=8CCWDwAAQBAJ

Nemitz P (2018) Constitutional democracy and technology in the age of artificial intelligence. Philos Trans R Soc, Math, Phys Eng Sci 376 (2133). https://doi.org/10.1098/rsta.2018.0089

Nielsen PA, Persson JS (2017) Useful business cases: value creation in IS projects. Eur J Inf Syst 26(1):66–83. https://doi.org/10.1057/s41303-016-0026-x

Norris P (2003) Preaching to the converted? Party Polit 9(1):21–45. https://doi.org/10.1177/135406880391003

Penz O, Sauer B (2019) Governing affects. https://doi.org/10.4324/9781351212434

Peters BG (2010) Bureaucracy and democracy. Public Organ Rev 10(3):209–222. https://doi.org/10.1007/s11115-010-0133-4

Poullet Y (2009) Data protection legislation: what is at stake for our society and democracy? Comput Law & Secur Rev 25(3):211–226. https://doi.org/10.1016/j.clsr.2009.03.008

Rheingold H (1994) The virtual community: finding connection in a computerized world. Harvill Secker, Minerva https://books.google.com/books/about/The_virtual_community.html?hl=&id=A6e27y–PlkC

Sklyar A, Kowalkowski C, Tronvoll B, Sörhammar D (2019) Organizing for digital servitization: a service ecosystem perspective. J Bus Res, February. https://doi.org/10.1016/j.jbusres.2019.02.012

Svensson L (2019) Tekniken Är Den Enkla Biten: Om Att Implementera Digital Automatisering i
   Handläggningen Av Försörjningsstöd. Socialhögskolan, Lunds Universitet, Research reports in
   social works
Vargo SL, Archpru Akaka M, Vaughan CM (2017) Conceptualizing value: a service-ecosystem
   view. J Creat Value 3(2):117–124. https://doi.org/10.1177/2394964317732861
Vargo SL, Lusch RF (2004) Evolving to a new dominant logic for marketing. J Mark 68(1):1–17.
   https://doi.org/10.1509/jmkg.68.1.1.24036
Vink J (2019) In/Visible—conceptualizing service ecosystem design. PhD thesis, Karlastad
   University Studies, 2019:17
VINNOVA (2018) Democracy data—innovation management-for a data smart social services.
   https://www.vinnova.se/en/p/democracy-data—innovation-management-for-a-data-smart-soc
   ial-services/
West DM (2004) E-government and the transformation of service delivery and citizen attitudes.
   Public Adm Rev 64(1):15–27. https://doi.org/10.1111/j.1540-6210.2004.00343.x
Wynn-Williams CE (1931) The use of thyratrons for high speed automatic counting physical
   phenomena. Proceedings of the royal society of London. Series A, containing Papers of a
   mathematical and physical character 132(819):295–310. https://doi.org/10.1098/rspa.1931.0102
Zittel T, Fuchs D (2006) Participatory democracy and political participation: can participatory
   engineering bring citizens Back in? Routledge, Abingdon

Petter Falk is a service designer at RISE—Research Institutes of Sweden and a Ph.D student tied
to political science and CTF—Center for Service Research at Karlstad University. In the intersection of political science, STS and critical studies Petter Falk is researching data as a democratic
artifact in the public sector. Working closely with actors in both health care and social services, his
research has explored the prerequisites and assumptions of data-culture through several research
and innovation projects, both in Sweden and at an international level.

Chapter 4
Innovation in Data Visualisation for Public Policy Making

Paolo Raineri and Francesco Molinari

Abstract In this contribution, we propose a reflection on the potential of data visualisation technologies for (informed) public policy making in a growingly complex and fast changing landscape—epitomized by the situation created after the outbreak of the Covid-19 pandemic. Based on the results of an online survey of more than 50 data scientists from all over the world, we highlight five application areas seeing the biggest needs for innovation according to the domain specialists. Our main argument is that we are facing a transformation of the business cases supporting the adoption and implementation of data visualisation methods and tools in government, which the conventional view of the value of Business Intelligence does not capture in full. Such evolution can drive a new wave of innovations that preserve (or restore) the human brain's centrality in a decision making environment that is increasingly dominated—for good and bad—by artificial intelligence. Citizen science, design thinking, and accountability are mentioned as triggers of civic engagement and participation that can bring a community of "knowledge intermediaries" into the daily discussion on data supported policy making.

Keywords Business intelligence · Technology innovation trends · Evidence-based policy making · Data scientist profession

4.1 Introduction: Data Visualisation Between Decision Support and Social Influence

Data visualisation is the art and science (Mahoney 2019) of graphically displaying large amounts of data in a visually attractive and simplified way, to facilitate understanding, decision and therefore action. This is done by a plethora of methods and

P. Raineri (✉)
Como, Italy

F. Molinari
Department of Architecture and Urban Studies, Politecnico di Milano, Milan, Italy
e-mail: mail@francescomolinari.it

© The Author(s) 2021
G. Concilio et al. (eds.), *The Data Shake*,
PoliMI SpringerBriefs,
https://doi.org/10.1007/978-3-030-63693-7_4

tools, many of which—such as pie charts, dashboards, diagrams, infographics and maps—are quite familiar to those who have even basic notions of statistics or simply follow the news on traditional and social media. In fact, popularisation of data visualisation is a now well established phenomenon, which roughly materialised in the beginning of the new century, when tag clouds began to show up on blogs and websites and the so-called sparklines—very small graphs embedded in lines of journalistic text, to show up trends and variations—were invented.[1]

The effectiveness of using images instead of (too many) words to describe data has been evident to researchers from many disciplines, including both natural and social sciences.[2] Even marketing—not to mention political communication—grasped the importance of visual displays to single out messages destined to be "digested" and transformed into actions by huge numbers of people, although sometimes at the cost of dissimulating, rather than refining, some true aspects of reality (Gonzalez 2019). In parallel, the so-called Business Intelligence field also took more and more benefit of visualisation technologies, especially with the growing size of data to be handled—both from within and outside the organisation—and the need to compress the decision making time of top and middle managers, by automating and simplifying the process of relevant information acquisition and analysis.

This peculiar aspect of data visualisation—being at the crossroad between decision support and social influence—has become particularly clear after the outbreak of the Covid-19 pandemic, when the first known cases of "deliberate censorship" have materialised on social media, such as Twitter and Facebook, to halt the spread of misinformation on how to protect against the virus. Not only are visuals now being used to place alerts on contested statements, but also the proliferation of infographics manipulating official data instrumentally has started to be cross-checked for reliability. In some extreme cases, Apple and Google are known to have removed some of these controversial apps from their stores (*Financial Express* 2020). On the other hand, the richness of mobility data as captured by the GPS of individual smart phones as well as the combination of textual contents with the geolocalisation of people interacting on social media have been widely perceived, maybe for the first time, as precious sources of information for decision making—and social influence again. Consider as examples: the use of Facebook and Google surveys done at Carnegie Mellon University to predict surges in the virus spread (Wired 2017); the Covid-19 Infodemics Observatory built at FBK in Trento[3] using a global dataset of tweets and GPS information; and the business alliance "for the common good" between Apple and Google to enable interoperability between Android and iOS devices and jointly develop a Bluetooth-based contact tracing platform (Apple Newsroom 2020). The latter has generated, among others, the "Immuni" mobile app that is now widely advertised by the Italian government as a form of prevention against the unwanted effects of the "next wave" of contagion (Reuters 2020).

---

[1] https://en.wikipedia.org/wiki/Sparkline.
[2] See: Data is beautiful: 10 of the best data visualization examples from history to today. https://www.tableau.com/learn/articles/best-beautiful-data-visualization-examples by Tableau Software (2020).
[3] https://covid19obs.fbk.eu/#/.

In this scenario, a crucial question to be posed to both researchers and practitioners of public administration, is whether we are facing the inauguration of a new trend for the take-up of data visualisation technologies in government. According to Fortune Business Insights (2020), the market of software applications for business intelligence and visual analytics, which nominally also includes public buyers, is estimated to hit \$19.2 billion in the next seven years, from the current \$8.85 billion, with an expected CAGR of 10.2% per. On the other hand, the pricing of business intelligence solutions is sometimes prohibitive, especially for small-sized public bodies and agencies, and statistics are missing on the impact of using open source solutions in the various application domains—such as public healthcare or urban planning. According to IDC (Shirer2019), the federal/central government share in the global market of business intelligence solutions is lower than 7% of total purchases. This figure either omits important buyers (e.g. local government or public utilities) and unpaid resources (such as free and open tools) or is simply an indication that the main business argument used to push adoption—"get to know more about what happens in your organisation, or just outside it, to take more informed decisions"—for a variety of reasons is not as compelling in the public sector as it seems to be for large corporations and medium sized enterprises.

This paper aims to stimulate a reflection in that direction, by asking the question of which kind of innovation is mostly needed to facilitate, rather than prevent, the take-up of data visualisation tools for public policy making. Answers to this question have been gathered from more than 50 domain experts (data scientists) from all over the world, by means of an online survey.[4] After elaborating on received answers, we contrast this sort of indirect collection of user requirements with other emerging or growingly established technology trends—including e.g. Artificial Intelligence, IoT (Internet of Things), Edge Computing and AR/VR (Augmented/Virtual Reality). Our conclusion is that innovation in data visualisation may contribute to preserve a sort of demilitarised zone, where human decisions prevail over machine intelligence and initiatives. This aspect should be particularly appreciated by policy makers, but is curiously not well developed by specialised software vendors.

---

[4]The survey was a poll with five questions. Participants could answer using a free text form. The poll was done with Google Forms[TM] and managed by the first author of this paper It was sent to members of the Data Visualization Society (https://www.datavisualizationsociety.com/). The audience was filtered before enabling access to the survey, to be sure about the participants' background. All participants have worked for a public institution at least once as data scientists or data visualization managers. 52 of them answered the survey. They were asked if they wished to appear as supporters of this study, 4 of them answered positively and are acknowledged here: Alessandro Chessa, Evgeny Klochikhin, Luca Naso and Sevinc Rende. The survey was open for one month, from February 9th until March 9th 2020. The five questions were: (1) What are the 3 most important troubles you face while doing data visualization for a policy maker or for the public sector in general? (2) What are the 3 most important rules you follow to deliver a data visualization that is really useful for your client? (3) What kind of "visualization modality" do you prefer to engage citizens in producing data and be aware of them? (4) In your opinion, what kinds of innovation in data visualization are the most viable and feasible for the next future? (5) Any articles or book suggestion to know more about this topic? Any "talks" we must listen to? The 210 answers were transcribed and clustered in macro-topics. All of them were useful to act as foundation of this contribution.

4.2 Scoping the Experiences of Data Scientists

Among the several definitions of innovation we adopt the one by Loughlan (2016): "the pursuit of a better service or product that adds value to organizations, communities and to the wider society". We asked domain experts from all over the world to help us define which value data visualisation tools (can) bring to public policy making. It turned out that there is not a single answer to such question. Value (to be) created depends very much on: (1) the policy maker's goals, (2) if and how they are communicated to the data scientist, and (3) whether the latter properly understood them or not. This doesn't mean that innovation is not part of the process—only that it must also involve the communication between interested parties. All professional data scientists agreed that the key rule to deliver an effective and useful visualisation for a policy making process is: know your "customer's"[5] goals. If the policy maker and the data scientists are not aligned on how the former will use the data provided by the latter, then the output of visualisation is at risk of being use-less.

As a complement to the above, it is relevant that data scientists identify the areas in which they see the biggest needs or pains from the perspective of their customers. In fact, one of the secure ways to create value with innovation is through tackling burning problems. The survey respondents mentioned the following issues:

- Multiple data source management: despite some recent progress in related technologies, data mash-up and cleaning are still the most time and resource consuming tasks of any data visualisation project. This is particularly true when multiple data sources are handled, lacking homogeneity and sometimes being non-standard, a frequent situation when working in/for the public sector.
- Rigorous data integration: the robustness and scientific lineage of the data used for policy making is vital. Unfortunately, it is still too difficult to certify and verify data sources and many steps forward should be done to avoid misinterpretation of what is being visualised. This pain has been reported by almost all of our survey respondents.
- Actionable information delivery: as obvious as it may be, data visualisation is supposed to generate ready-made insights that will lead the policy maker to a decision, speeding up the whole process and fostering stakeholders' engagement. However, this is not always the case and of course, value creation is correspondingly reduced.
- Personalised user experience: the way different people look at a same dashboard or chart can be quite different. Yet as we mentioned above, data visualisation has (or should have) the capacity to trigger human brain to a specific decision and (re)action. In this perspective, we still know (and practice) too little on how to adjust the user experience according to the various personal behaviours while enjoying data.

---

[5]The word "customer" in the present document is referred to policy makers being customers of data visualization.

These issues have been reported by more than 50 data visualisation experts, working with policy makers or public institutions in different countries. However, their relevance for the state of the art is also confirmed by the literature and personal experience highlights we are going to present below.

4.2.1 Multiple Data Source Management

Various start-ups are tackling the issue of merging multiple data sources, but none of them seems to have reached the "nirvana" for the average data scientist. This is especially true for the non-relational sources. The main reason why the problem is not so easy to address is the variety of potential applications. Obviously, every process has its different goals, so the quality of your mash-up relies on what you should use those data for. This has been documented by Samuelsen, Chen and Wasson (2019) in a literature search of more than 115 publications, only restricted to the learning analytics domain. There are other reasons, however, which transform this issue into one of the hardest technical pains of this profession. Even if you use one of the good data wrangling tools now on the market, you still need at least some python/R basics or an old school manual on Excel to come up with a decent result. This means to allocate hundreds of working hours to simply get ready, instead of delivering the analytics and visualisation outputs.

4.2.2 Rigorous Data Integration

This pain is both a "call for better data" and a request for improving their informative value. William Davies wrote a very good article in 2017 (Davies 2017) that still represents a good starting point for a crucial discussion to all of us. Getting scientifically validated data should always be a main concern for every policy maker, but when it comes to being sure about data lineage and research methods your legs start to crumble. Open data might be an alternative, but some problems remain: "1) Data is hard (or even impossible) to find online, 2) data is often not readily usable, 3) open licensing is rare practice and jeopardized by a lack of standards" (Lämmerhirt et al. 2017).

Data scientists also want to be sure that the visualisation output is not misleading in any way. In some cases, some colour palette mistakes, or naive data manipulations in the analytical steps, may lead to that very risky outcome, since the policy makers might ground their decision on a false representation of facts. In the future, Artificial Intelligence may help us solve this very tiny but insidious problem. Just as we now have language spell checkers for our weird grammar typos, some advisor bots may soon help us remember what we did with data and where it comes from, before taking any decision based on it.

4.2.3 Actionable Information Delivery

Delivering an actionable output is in the top five rules of data scientists since Florence Nightingale taught to the world what the superpowers of data visualisation were meant for. She, as a nurse, drew herself data charts to boost decision making for the British army recovery in the Crimean war (Kopf 1916). But it's not that easy. The very meaning of the word "actionable" puts a responsibility on data visualisation creators, especially within the domain of policy making. The challenge as we mentioned already, is to fulfil the policy maker's goals while at the same time empowering him/her in a way that shortens the decision process time. Artificial Intelligence and particularly Natural Language Processing techniques are being trialled as alternative solutions for an actionable data visualisation. However, they still take too much processing time, resulting in a boring user experience and a poor quality of the generated insights.

4.2.4 Personalised User Experience

Several authors (Toffler 1970; Davis 1987; Womack 1993; Anderson and Pine 1997) introduced the concept of mass customization as the "next big thing" in modern manufacturing. It took us maybe 30 years to reach a point of no return but since the 2000s it has become clear that the concept fits perfectly into actual human needs. We leave the moral question to other occasions. Here we simply note that a similar need is felt in the data visualisation for policy making community, as the responses to our survey demonstrated. We can imagine a visualisation able to adapt its colours, shapes, space, insights to personal behaviours and preferences. The fast-paced innovations of Machine Learning and more broadly Artificial Intelligence (also including mixed reality and face/voice recognition) are natural candidates to fulfil this requirement. This is probably the most attended innovation in data visualisation now, so we can hope to see it as a reality sooner than we think.

4.3 A Critical Eye on Technology Innovation Trends

The oft-cited Artificial Intelligence is not the only route of innovation that can push up the threshold of data visualisation technologies in support of public decision making. Internet of things (IoT) (Sethi and Sarangi 2017) together with the new Edge Computing wave—the calculation model in which data is processed by the device itself or by a local computer or server, rather than being transmitted to a data centre (Premsankar et al. 2018), as well as voice/image recognition are also worth consideration.

IoT is the enabler of the "sensibility" of a country, region or city. We can say that IoT sensors act for a city just like the human receptors act for our body sensibility. That's why MIT started to name smart city topics as the business of "sensible cities" (Dizikes 2016). The role of data visualisation in this context is quite obvious. The policy maker should merge sensible city projects, predictive analytics, and data visualisation to be able to act as the "wisdom brain" for good decisions.

Edge Computing is a distributed computing paradigm that brings computation and data storage closer to the location where it is needed, to improve response times and save bandwidth. This topic comes with high relevance in this paper because it's already present in real world and only needs to be embraced to start producing effects. Endowed with this "data wisdom brain" the policy maker would boost exponentially his/her odds of success in every decision. Although such thoughts might lead to an enormous discussion about the future of humanity as a whole (Harari 2016) we could try to stay humble and admit that a more informed decision is always a better decision. The more data and information you have with you, the better your choices will be. Our assumptions and beliefs about the importance of this topic will become obvious if you agree with this sentence.

The same computational power needed to let IoT and predictive analytics play an effective role in decision making can also enable language-related and image-related technologies. There are many implications of this field. Voice recognition enables hands off interaction with machines. Natural language processing allows to understand human language shades and return warmer outputs. Image recognition enables to detect human emotions. The most famous implication of this kind of technologies is represented by the deep fake world (Vincent 2018). Here fake moving images of famous personalities are created leading the audience to believe in some weird videos (many about the presidents of USA, Russia, North Korea, Germany went viral in the social media just a few years ago). Notwithstanding the bad fame due to the heavy privacy implications, if a policy maker started to use these solutions then a new generation of data visualisation tools would help tremendously improve the engagement and truthfulness levels of the policy making cycle. This because that kind of technologies would speed up the creation of a lot of informative content and material and boost the engagement rate of the target audience thanks to a super personalized and customer-centric communication.

However, this potential does not seem to be perceived as such in the public policy making community. Let's take another field as a benchmark case to machine learning, namely the professional basketball community. Not many years ago, M.I. Jordan commented that despite the diffused awareness of the importance of data analytics and therefore visualisation, *"we are no further ahead than we were with physics when Isaac Newton sat under his apple tree"* (Gomes 2014). And yet in the basketball community such knowledge gap was filled in by a single, although enthusiast, student of engineering with the simple (but brilliant) introduction of SOM techniques (Kohonen 1982, 2001) into players' analytics (Bianchi et al. 2017). What can be the equivalent of that "connecting the dots" innovation in the policy making field? We have two or three possible ideas in mind.

A serious candidate is civic engagement. The ability to promote active interactions with citizens, not only as consumers but also producers of data, is nowadays well accepted as a wise and powerful way to procure useful information for public policy making. Getting granular, rigorously gathered data from a number of collaborative citizens is commonly called citizen science (Hand 2010; Castelvecchi 2016). This great way of engaging people with institutions has been used for an amazingly wide set of topics. But what is the link with data visualisation? Grasping the full potential of citizen science basically relies on people's understanding of the data they collect. A great example is the CIESM JellyWatch project, a citizen science survey born after an overall jellyfish review in 2013 (Boero 2013) where a citizens mobile app enabled the collection of an enormous amount of data about Mediterranean Sea jellyfish distribution (Marshall 2010). Though bringing enormous benefits, citizen science must be managed in a good manner to avoid its risks. A good reading about the pros and cons of this approach is the article on Nature by Aisling Irwin (2018).

Another good candidate is the introduction of "design thinking" methodologies in the data visualisation journey. Design thinking is an umbrella term for the cognitive, strategic, and practical processes by which design concepts are developed. Many of the key processes of design thinking have been identified through studies, across different design domains, of cognition and activity in both laboratory and natural contexts. Design thinking is a way to put the end-user at the core of the design process. This does not only result in a faster and more effective output delivery for the policy maker, but introduces many connectors to the topic of citizens' engagement. If policy makers would like to engage citizens nowadays, they should always look for a participation trigger. Data availability (and open data) draws an honest and transparent pathway that always acts as a nudge towards citizens' engagement (citizen science plays a queen role in this game). Another nudge is to build the whole project with a "design thinking" vision. Citizens' problems, their educational levels, interests, etc., everything should be taken into account. Communication with the audience should be tailored, direct and tackling only the main issues, avoiding the exchange of useless information. Following this train of logic, every data visualisation would appear familiar to the citizens, something made for them. However, in this quest for engagement "devil is in the details". Both security and data quality issues play a major role in making this engagement pathway really workable for the policy maker who wants to benefit from the data visualisation features. The amazing work made by some projects in this area is already on the market waiting to be leveraged by public institutions (see Wired 2017; Kambatla et al. 2014). This is coherent with the digital transformation the whole world is having towards a human-centred approach to innovation (Kolko 2015).

Finally, choosing to visualise policy relevant data in a way that people both enjoy and understand is the perfect "final step" of an accountability process. Having a good and effective data visualisation leads to many positive implications: it highlights what is relevant and avoids distractions; proves the decision outcomes; helps to stay focused on budget and efforts; inspires hands-on participation; nurtures effective communication; flattens the learning curve on how to visualise data for decision making.

From a technological point of view there are some known tricks to make sure that data visualisation strikes the goals of civic engagement. The easiest way is always to start by keeping in mind the pre-attentive attributes like colours, shapes, movement, spatial positions etc. (Ware 2004). This could appear as a small issue, or a trivial aspect. But the more you push data visualisation forward, using it for real decision making, the more does this aspect become crucial, marking the difference between a good or a bad policy decision. Take the following example: you are in 2025 producing an important 3D visualisation directly going into the smart glasses of your citizens and forgot to think about colour-blind people. What could be the fallbacks? The number of delivered contents is also very important. This should be limited to those you want your citizens to follow. It is recommended giving to the users the possibility of drilling down and zooming in if they want, but the first look and feel must be lean and essential. Finally, it is important to always give to people the possibility of getting a "multi-dimensional" exploration of data. Maps are the best way to deliver a content in this way (and many of our survey contributors confirmed that).

4.4 Conclusions and Way Forward

Be it because of the pains highlighted in our survey of domain experts or the technical limitations of many software applications, policy makers around the world do not seem to be ready yet to adopt data visualisation tools in support to decision making. And the main argument conventionally used to promote Business Intelligence in the private sector—that evidence is key to take informed decisions (Davies et al. 2000)—does not seem to work as well here.

In the previous section we have pointed at citizen science, design thinking and accountability as three triggers of civic engagement and participation that can bring a community of "knowledge intermediaries" into the daily discussion on policy making (Isett and Hicks 2019). This can help push that community ahead: from passively knowing the theory of a thing, to taking active action to carry it forward. But there is more: this evolution can drive a new wave of innovations preserving (or restoring) the human brain's centrality in a decision making environment that is increasingly dominated—for good and bad—by Artificial Intelligence.

In fact, looking at "the big picture" it becomes clear that the ultimate goal (or outcome) of Artificial Intelligence is to prevent the beneficiaries of data visualisation from interacting directly with data. What we propose to do instead is to create visualisations that respect some particular constraints we gave them previously. Then we should look at the charts and try to infer a decision. By doing so we would always inject our human "bias" (for good and bad) into all the steps from database querying to the final chart design. Are we sure that we really want to get rid of this? Again, the answer may not be that straightforward.

Struggling with complexity may be a good argument in favour. As shown on the occasion of the Covid-19 health and information pandemics, there is no simple way other than visualisation to do justice of zillions of data growing in real time at an unprecedented speed. There is a large world of analytics and representation techniques to fight with complexity. The ability to process a big amount of data in various formats, from various sources, and deliver meaningful information to forecast future outputs to users is known as predictive analytics. Predictive analytics impacts almost every domain (Wang et al. 2018) and should be viewed as the main compass of every policy maker. The evolution of complex network analysis will deeply contribute to this area in helping to see both the big picture of a complex system and highlighting its peculiarities. We see this as the best way to be able to zoom in deeply to look for specific issues and solutions. Generally speaking, big data analytics open a bright future for our economies and societies (Amalina et al. 2020).

Keeping the control of our destinies—including the possibility of making mistakes—is a good reason against. Talking about the necessity of staying up to date, it is impossible to argue against the fact that nowadays we are all under pressure. Every public and private actor is somehow in a rush for the, so called, change. There is a shared pain about staying aligned with the world's pace and being able to move fast and in the right direction at the same time. This feeling is even heavier for the policy makers that guard the keys of our world. But is this enough to decide that artificial intelligence should take full control?

Luckily, data science has evolved in parallel with the raising amount of data we produce every day. Although data scientists agree that the data amount grows faster than our ability to analyse them, we can say that "the challenge" is still fair enough. Therefore, leaving philosophical worries behind us, what we suggest to the data visualisation managers in the policy making area is to undertake an open and wide investigation to bring existing innovations from other sectors. Many great things are now ready to be implemented, coming from unexpected areas. Innovation in geo-spatial analytics developed for basketball could also be useful for smart mobility (Metulini et al. 2017). Old scientific research might lead to a solution for an institution's data merging process (Buja et al. 1996). New methods and bio technologies (PHATE: Potential of Heat Diffusion for Affinity-based Transition Embedding) for visualising high-dimensional data (Moon et al. 2019) could revolutionize some aspects of the policy making cycle. Young students out in the world playing with data and "accidentally" solving many smart cities issues (Yang et al. 2019) could be scouted to speed up the innovation routines. Scientific projects already tailored for policy makers (Tachet et al. 2017) could be investigated with less fear. Future applications of mixed (i.e. augmented + virtual) reality (Joshi 2019) will be involved for sure in the next steps of data visualisation for policy making, with a particular attention to citizens' engagement. Immersive videos, educational classrooms, policy maker meetings, political surveys, interactive discussions—the potentialities of mixed reality merged with data visualisations are infinite.

In conclusion, we can say that, once again in history, what will make the difference between remaining "stuck in the present" and evolving to a bright future will be the ability to contaminate mindsets and cross-fertilise domains, bearing in mind what really matters for people and avoiding to adhere to the "coolness-driven" purchasing decisions.

Acknowledgements We thank Francesca Montemagno for her support, Prof. Grazia Concilio for this opportunity and all the data visualization innovation survey participants for their precious inputs.

References

Anderson D, Pine J (1997) Agile product development for mass customization: how to develop and deliver products for mass customization, niche markets, JIT, build-to-order, and flexible manufacturing. Chap 1:2, 3–39, Irwin Professional Pub, Chicago

Amalina F, Hashem IAT, Azizul ZH, Fong AT, Imran M, Anuar ANB (2020) Blending big data analytics: review on challenges and a recent study. IEEE Access 8:3629–3645

Apple Newsroom (2020) Apple and Google partner on Covid-19 contact tracing technology. 10 April. https://www.apple.com/newsroom/2020/04/apple-and-google-partner-on-covid-19-contact-tracing-technology/. Accessed on September 2020

Bianchi F, Facchinetti T, Zuccolotto P (2017) Role revolution: towards a new meaning of positions in basketball. Electron J Appl Stat Anal 10(3):712–734

Buja A, Cook D, Swayne DF (1996) Interactive high-dimensional data visualization. J Comput Graph Stat 5(1):78–99. https://doi.org/10.1080/10618600.1996.10474696

Castelvecchi D (2016) Citizen scientists take on latest gravitational-wave data. Nature. 9 March. https://www.nature.com/news/citizen-scientists-take-on-latest-gravitational-wave-data-1.19505. Accessed on September 2020

Boero F (2013) Review of jellyfish blooms in the mediterranean and black sea. GFCM Stud Rev 92

Davies HTO, Nutley SM, Smith PC (2000) What works? evidence-based policy and practice in public services. Chap. 1:2:16, Bristol, Policy Press

Davies W (2017) How statistics lost their power—and why we should fear what comes next. The Guardian, 19 January. https://www.theguardian.com/politics/2017/jan/19/crisis-of-statistics-big-data-democracy. Accessed on September 2020

Davis S (1987) Future perfect. Addison-Wesley Publishing, New York

Dizikes P (2016) Senseable city lab: cities of tomorrow. MIT News Office. http://news.mit.edu/2016/book-cities-tomorrow-urban-design-0705

Financial Express (2020) Apple, Google join Twitter and Facebook in war against fake news on Coronavirus. 6 March. https://www.financialexpress.com/industry/technology/apple-google-join-twitter-and-facebook-in-war-against-fake-news-on-coronavirus/1891062/. Accessed on September 2020

Fortune Business Insights (2020) Data visualization market size, share and industry analysis, by component (software, services), by solution (standalone visualization software, embedded data visualization module), by enterprise size (small enterprises, medium enterprises, and large enterprises), by industry (BFSI, construction and real estate, consumer goods, education, government, healthcare and pharmaceuticals, information technology, services) and geography forecast, 2020–2027. https://www.fortunebusinessinsights.com/data-visualization-market-103259. Accessed on September 2020

Gomes L (2014) Machine-learning Maestro Michael Jordan on the delusions of big data and other huge engineering efforts. https://spectrum.ieee.org/artificial-intelligence/machine-learning/machinelearning-maestro-michael-jordan-on-the-delusions-of-big-data-and-other-huge-engineering-efforts. Accessed on September 2020

Gonzalez XV (2019) Honest visuals: ethics in data visualisation. Keynote speech de-livered at the European Commission's conference "EU Data Viz", Luxembourg, 12 November. https://tinyurl.com/yxfpvmb8

Hand E (2010) Citizen science: people power. Nature 466:685–687. https://doi.org/10.1038/466685a

Harari YN (2016) Homo deus: a brief history of tomorrow. Harvill Secker, London

Shirer M (2019) IDC forecasts revenues for big data and business analytics solutions will reach $189.1 billion this year with double-digit annual growth through 2022. 4 April. https://www.idc.com/getdoc.jsp?containerId=prUS44998419. Accessed on September 2020

Irwin, A. (2018) No PhDs needed: how citizen science is transforming research. Nature, 23 October. https://www.nature.com/articles/d41586-018-07106-5. Accessed on September 2020

Isett KR, Hicks K (2019) Pathways from research into public decision making: intermediaries as the third community. Perspect Public Manag GovAnce 3(1):45–58. https://doi.org/10.1093/ppmgov/gvz020

Joshi N (2019) 8 future mixed reality applications to watch out for. Forbes, 3 November https://www.forbes.com/sites/cognitiveworld/2019/11/03/8-future-mixed-reality-applications-to-watch-out-for/#1ec5d7b33465. Accessed on September 2020

Kohonen T (1982) Self-organized formation of topologically correct feature maps. Biol Cybern 43:59–69

Kohonen T (2001) Self-organizing maps, 3rd Extended edn. Springer Series in Information Sciences, vol 30. Springer-Verlag, Berlin, Germany. ISBN 978-3-540-67921-9

Kambatla K, Kollias G, Kumar V, Grama A (2014) Trends in big data analytics. J Parallel Distrib Comput 74(7):2561–2573

Kolko J (2015) Design thinking comes of age. Harvard Business Review. September 2015 Issue https://hbr.org/2015/09/design-thinking-comes-of-age

Kopf EW (1916) Florence nightingale as statistician. Publ Am Stat Assoc 15(116):388–404

Lämmerhirt D, Rubinstein M, Montiel O (2017) The state of open government data in 2017. Open Knowledge International. June. https://blog.okfn.org/files/2017/06/FinalreportTheStateofOpenGovernmentDatain2017.pdf

Loughlan C (2016) The innovation manifesto. Cambridge Institute for Innovation

Mahoney M (2019) The art and science of data visualization. Medium, 14 October. https://towardsdatascience.com/the-art-and-science-of-data-visualization-6f9d706d673e. Accessed on September 2020

Marshall A (2010) Stinging season: Can we learn to love the jellyfish? Time, 25 August. http://content.time.com/time/article/0,8599,2012178,00.html

Metulini R, Manisera M, Zuccolotto P (2017) Space-time analysis of movements in basketball. In: Pertucci A, Verde E (eds) Proceedings of the international meeting of Italian statistical society, "Statistics and Data Science: new challenges, new generations" Florence, Italy, 28–30 June 2017

Moon KR, Van Dijk D, Wang Z et al. (2019) Visualizing structure and transitions in high-dimensional biological data. Nat Biotechnol 37:1482–1492. https://doi.org/10.1038/s41587-019-0336-3

Premsankar G, Di Francesco M, Taleb T (2018) Edge computing for the internet of things: a case study. IEEE Internet Things J 5(2):1275–1284

Reuters (2020) Italians embrace coronavirus tracing app as privacy fears ease. 11 June. https://www.reuters.com/article/us-health-coronavirus-italy-apps/italians-embrace-coronavirus-tracing-app-as-privacy-fears-ease-idUSKBN23I2M5. Accessed on September 2020

Samuelsen J, Chen W, Wasson B (2019) Integrating multiple data sources for learning analytics review of literature. Res Pract Technol Enhanc Learn 14(11). https://doi.org/10.1186/s41039-019-0105-4

Sethi P, Sarangi S (2017) Internet of things: architectures, protocols, and applications. J Electr Comput Eng 2017 lArticle ID 9324035

Tachet R, Sagarra O, Santi P, Resta G, Szell M, Strogatz S, Ratti C (2017) Scaling law of urban ride sharing. Scientific Reports 7, 42868. https://doi.org/10.1038/srep42868

Toffler A (1970) Future shock. Random House, New York

Vincent J (2018) China's state-run press agency has created an 'AI anchor' to read the news. The Verge. 8 November, 5:15am EST

Wang Y, Kung L, Byrd TA (2018) Big data analytics: understanding its capabilities and potential benefits for healthcare organizations. Technol Forecast Soc Chang 126:3–13

Ware C (2004) Information visualization: perception for design. Morgan Kaufmann, 2nd edn. San Francisco, CA

Wired (2017) The web's greatest minds explain how we can fix the internet. Wired, 20 December. https://www.wired.co.uk/article/the-webs-greatest-minds-on-how-to-fix-it

Womack JP (1993) Mass customization: the new frontier in business competition. MIT Sloan Manag Rev 34(3):121

Yang C, Zhang Y, Tang B, Zhu M (2019) Vaite: A visualization-assisted interactive big urban trajectory data exploration system. 2019 IEEE 35th international conference on Data Engineering (ICDE), Macao: 2036–2039

Paolo Raineri Digital Innovation Expert and Data-driven advisor. Paolo brought his scientific background in marine biology into the entrepreneurial world starting his own company in 2010 tackling big data sports analytics issues in basketball. The work he did with his company has been recognized by various academies, such as MIT, Lewis University, Michigan state university, Politecnico di Milano, University of Pavia. He has been involved in different public sector conferences and workshops (MIO ECSDE; ACR+; UNEP/MAP; WCMB; CIESM congress …) in which he had the chance to work towards a better communication of results and a better visualization and storytelling of environmental topics. During the last decade he dedicated his efforts in various data-driven topics trying to merge the digital transformation world with the SMEs and Public sector issues. Lately he also worked for the very first Open Source BI solution ever born. He's now a digital consultant.

Francesco Molinari is an international researcher and policy advisor with a 20-years working experience in R&D and innovation projects and programmes at European, national and regional levels. Formerly he has been engaged for about 12 years in territorial marketing and the delivery of financial services to SMEs—including support to EU grants access for their green and brown field investments. For 5 years he has served in a top managerial position at a middle-sized Municipality.

Part II
The PoliVisu Project

Chapter 5
Policy-Related Decision Making in a Smart City Context: The PoliVisu Approach

Yannis Charalabidis

Abstract Dealing with the growing quest for better governance, the advancement of ICT provides new methods and tools to politicians and their cabinets on an almost daily basis. In this changing landscape, the PoliVisu project constitutes a step forward from the evidence-based decision making, going towards an experimental approach supported by the large variety of available data sets. Through utilizing advanced data gathering, processing and visualisation techniques, the PoliVisu platform is one of the most recent integrated examples promoting the experimental dimension of policy making at a municipal and regional level.

Keywords Policy-related decision making · Evidence based policy making · Experimental policies

5.1 Evidence-Based Policy Making and the Rise of ICT

The need for utilizing advanced Information and Communication Technologies (ICT) infrastructures and services, for assisting public sector decision makers in reaching justified decisions has been a primary role for technology in our society, since the previous century. Coined as "evidence-based policy making" in the UK and rapidly expanding to the US and the rest of the world since the rise of the 21st century (Sanderson 2002), this attempt to follow the practices of science in public management has found significant interest among Digital Governance scholars, practitioners and communities.

In this quest for better governance, the advancement of ICT provides new methods and tools to politicians and their cabinets on an almost daily basis. Some of the key technological offerings to support policy making in a systematic manner but not always in an easy way, are the following:

Y. Charalabidis (✉)
University of the Aegean, Samos, Greece
e-mail: yannisx@aegean.gr

© The Author(s) 2021
G. Concilio et al. (eds.), *The Data Shake*,
PoliMI SpringerBriefs,
https://doi.org/10.1007/978-3-030-63693-7_5

- The vast amounts of data that can now be acquired, managed, stored and reused forming what we now call Big, Open and Linked Data (BOLD)—a basic layer for the myriad of processing tools to follow (Janssen et al. 2012).
- The development of systems and services for enabling citizen participation in various parts of the policy making cycle. Systems of e-participation, e-deliberation, even e-voting and collaborative design, are giving citizens more opportunities to take a vivid part in decision making at local, national, or international level.
- The rapid advancement in systems and services making use of artificial intelligence (AI), that offer novel opportunities to understand societal phenomena, make advanced simulations to analyse and predict the impact of policy decisions—while also making decision making faster but also less transparent in some cases (Androutsopoulou et al. 2019).
- The evolution of data visualisation and visual analytics toolsets, providing ways to give new meaning to numbers and more levels of abstraction that can support the capabilities of the specialists but also attract the attention of the non-experts (Osimo et al. 2010).

5.2   ICT-Enabled Policy Making in a Smart City Context

As technological evolutions in the areas of Big, Open and Linked Data, Artificial Intelligence, Visual Analytics and Electronic Participation platforms are getting more connected to policy making, new opportunities arise for policy makers at all levels of administration. Through multi-method applications, ICT can now assist the public sector at all levels to untap completely new opportunities such as:

- Identification of possible policy interventions, through combined big data analytics and citizen participation with advanced opinion mining.
- Ex-ante policy impact assessment, through data analysis and societal simulation, combining techno-economical with behavioural parameters.
- Ex-post impact assessment of policy and legislation, integrating the monitoring of myriads of sensor-based indications together with citizens' sentiment analysis on policy measures and laws.
- Advanced monitoring of societal evolution in relevant policy areas, through complex dashboards supporting better public policy making and even legislation preparation at real time.
- Real-time decision making, through the application of advanced algorithms making use of openly available big data, with the proper regulation for transparency and good governance.

In parallel with increased ICT utilization in policy making, another big move in societies is happening: Urbanisation and the need for stronger local governance and advanced services provision gives rise to the Smart Cities movement, where policy making is moved from the national level to regions and municipalities who now also require the advanced ICT services at equal to the central government levels (Albino et al. 2015).

Along these two major axes of ICT utilization in policy making and the empowerment of municipal and regional administration that leads to Smart Cities, is where the PoliVisu project on "Policy Development based on Advanced Geospatial Data Analytics and Visualisation" takes place.

The PoliVisu project constitutes a step forward from the evidence-based decision making, going towards an experimental approach supported by the large variety of available data sets. Through utilizing advanced data gathering, processing and visualisation techniques, the PoliVisu platform is one of the most recent integrated examples promoting the experimental dimension of policy making at a municipal and regional level. As illustrated in Fig. 5.1, the PoliVisu approach combines data openness and citizen participation with advanced use of ICT and big data utilization, clearly differentiating from previous paradigms, like:

- Traditional e-participation systems, where citizen involvement may be great (if the system is properly used and populated) but use of ICT in policy making is minimum.
- Advanced big-data based analytics, where the amount of data, the complexity of processing and the validity of results for policy makers may be high, but citizen involvement and "buy-in" usually suffer.

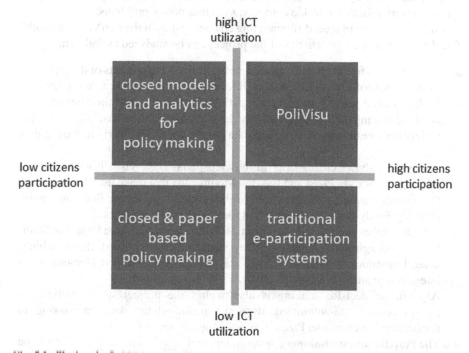

Fig. 5.1 Placing the PoliVisu approach in the collaborative decision-support map

- Traditional, not open or collaborative ways of decision making, where both openness and performance are a significant question, that nevertheless are still in operation in not a small minority of cities over the world.

5.3   The Unique Characteristics of the PoliVisu Approach

PoliVisu is a Horizon 2020 Research and Innovation (R&I) project with an aim to improve the traditional public policy making cycle, using big data and geospatial information visualisation techniques. The broad objective of the project is thus to assist public sector decision making at city level to become more open and collaborative by experimenting with different policy options through impact visualisation and by using the resulting visualisations to engage and harness the collective intelligence of policy stakeholders towards the development of collaborative solutions (PoliVisu 2020).

Working with real problems from three cities and a region (Issy-les-Moulineaux in France, Plzen in Czechia, Ghent and Flanders Region in Belgium) to address societal problems linked to smart mobility and urban planning, PoliVisu vision is to enable public administrations to respond to urban challenges by enriching the policy making process with tools for policy experimentation. The project supports three different steps of the policy cycle (design, implementation, and evaluation) in an attempt to enable the city officials to tackle complex, systemic policy problems.

But which are the unique, differentiating characteristics of the PoliVisu approach? The differentiating characteristics of the project can be analysed as following:

a. PoliVisu tools for urban and traffic planning utilize vast amounts of data, coming from a variety of sources like sensors, information systems and citizen inputs (e.g. Traffic counts, public transport data, parking availability, real time bus tracking and bike sharing data, as utilized in the Issy-les-Moulineaux case). The volume, variety and continuous flow of such data put the approach clearly in the big data area.

b. The project shows clear merits in the Geospatial Data Visualisation, through pilot applications in Plzen and Issy-les-Moulineaux that make use of active maps that visualise traffic volumes, public works planned, passenger flows and more, allowing for dynamic monitoring and routing of traffic volumes.

c. PoliVisu makes an important contribution in the area of real-time Dynamic Dashboards, through the PoliVisuals policy visualisation dashboard, that combines several monitoring, visualisation and real-time decision support elements in an integrated manner (PoliVisuals 2020).

d. Algorithmic decision making is also within the project scope, utilized in automated or semi-automated, dynamic, traffic-related decision making in applications in cities like Plzen, Issy-les-Moulineaux or Ghent.

e. The Polivisu approach shows a deep understanding of issues and problems to be tackled at Municipal and Regional Level, where policy making needs to be more pragmatic and results-oriented but where cross-city collaboration and fundraising for ICT infrastructures can be extremely challenging.

f. Finally, the project shows novel approaches in the areas of citizen participation and collaboration via digital means, where citizens provide data inputs, see and reuse openly available information and can also contribute to problem solving in a crowdsourcing way.

The above characteristics of the PoliVisu approach are illustrated in Fig. 5.2, where the project offering is compared to two others, well defined and currently often utilized systems:

- A traditional e-participation system, where citizen participation and municipal/regional orientation are emphasized over data acquisition, processing and visualisation that are usually weaker.
- A traditional Geographical Information System (GIS) with advanced analytics and capabilities in traffic management and urban planning, but where openness and collaboration with citizens, businesses and other authorities remain in question.

The comparative analysis shows the merits of the approach, that combines advanced data analytics, geospatial data processing and visualisation, with citizen engagement and collaboration at municipality or regional level.

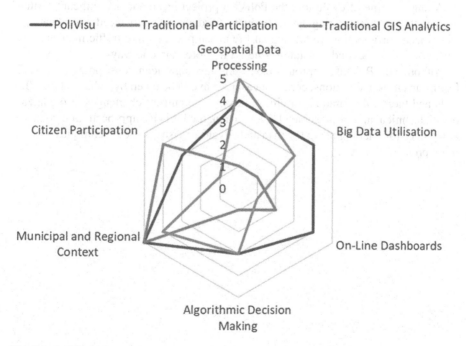

Fig. 5.2 PoliVisu compared to traditional e-Participation or GIS Analytics systems

5.4 Barriers and Limitations to the Full Exploitation of Data Potential in Policy Making

All the above being said on the merits of utilizing ICT in public decision making, and also the various PoliVisu achievements, ICT-enabled, data-driven policy making is far from being a "rose-garden". A number of barriers that prevent the use of ICT in policy making from becoming mainstream have been identified over the last decade, at least (Oliver et al. 2015; United Nations 2020). A brief analysis of those barriers and challenges is depicted in Table 5.1

5.5 Conclusion

All indications from research and practice, as well as the rapid technological evolutions in the areas of Big Data, Internet of Things, and Artificial Intelligence show that experiments-based policy making, or ICT-enabled decision support in the public sector, is a real need in public governance at different institutional levels (municipal, regional and national) in response to the growing request for transparency of public decisions.

Along this line of evolution, the PoliVisu project has made a significant contribution, through integrating large amounts of data with advanced visualisations and citizen participation, to tackle real-life urban planning and traffic management problems, going beyond the state-of-the-art in more than one ways.

Although the PoliVisu approach and results have significant reuse potential among European cities and regions, certain measures have to be taken by public sector officials and their collaborators, in order to overcome current challenges at organizational, technical and event societal levels. Then, the PoliVisu approach for evidence-based decision making using big data and advanced visualisation techniques will be more prone to success.

Table 5.1 Barriers for data-driven decision making

Barrier/Challenge	Description
Skills of Policy makers	Policy makers and decision makers (such as Ministers, Mayors, Region directors or other senior officials with decision making roles) should be able to understand and interpret reports in data analytics for value-adding insights and decision making while also being able to generate desired outcomes and impacts through strategic decision making. These new skills for policy makers may be even more difficult to become the mainstream, at local and regional level
Capacity and Interoperability of ICT tools and algorithms	Although technical barriers (e.g. the capacity of tools to assist in tackling a complex issue) are sooner or later being overcome by the rapid technological evolution, there are some aspects of the needed infrastructure that are still widely unavailable: (a) the ability of software models to analyse the non-techno/economical, behavioural aspects of societal problems or (b) the interoperability elements that would make such tools easily interconnected to one-another or (c) the mere capacity of such software models to understand and simulate situations of extreme complexity are still a quest and not a standard
Governance of Personal Data	Since most of the real-life applications of data-driven decision making involve the acquisition, processing, storage or publication of information that contains personal data of the citizens, a relevant regulatory framework has to be in place (aka in Law), so that both citizens and public sector officials feel adequately secure with such approaches. For local and regional administrations, this can be an even more high barrier, as such organisations typically cannot develop and enforce such a regulatory framework themselves, but have to wait for solutions at national level

(continued)

Table 5.1 (continued)

Barrier/Challenge	Description
Intension and Vision of Policy Makers	For experiments-based policy making attempts to turn finally successful, the high-level public sector officials (e.g. Ministers, secretaries, directors general, or other senior officials) must have a long-term vision for transforming policy making. This vision must be able to overcome or "absorb" the possible shortcomings or failures that will appear on the way. For the vision to be strong enough, an underlying intention to allow the "machine" to propose or identify solutions—against the human will sometimes, have to be present
Skills of Researchers	The researchers and practitioners that are engaged in data-driven policy making experiments must have a "multi-faceted" collection of skills: they have to be trained academically and have specific technical skills (e.g. able to deal with Python and other data tools or able to handle database infrastructure, data warehousing and statistics) while also they must have a non-trivial contextual understanding of the domain and the decision-making environment (e.g. knowledge of the city context and the specific problems with citizen mobility)
Collaboration with the Private Sector	Partnerships constitute an essential component of the data ecosystem for public decision making: a collaborative configuration involves the Government providing opportunities for public and private actors, that drive data innovation for the creation or modification of e-services with the aim of increasing economic or social benefits or otherwise generating public value. Enabling and empowering data-driven decision making infrastructures and services, involves making data widely available and creating opportunities for organisations and businesses to leverage on them

References

Albino V, Berardi U, Dangelico RM (2015) Smart cities: definitions, dimensions, performance, and initiatives. J Urban Technol 22(1)

Androutsopoulou A, Karacapilidis N, Loukis E, Charalabidis Y (2019) Transforming the communication between citizens and government through AI-guided chatbots. GovMent Inf Q 36(2):358–367

Janssen M, Charalabidis Y, Zuiderwijk A (2012) Benefits, adoption barriers and myths of open data and open government. Inf Syst Manag 29(4):258–268

Oliver K, Lorenc T, Innvær S (2015) New directions in evidence-based policy research: a critical analysis of the literature. Health Res Policy Syst 12(34)

Osimo D, Lampathaki F, Charalabidis Y (2010) Policy-making in a complex World: can visual analytics help? In: Kohlhammer J, Keim D. (eds) Proceedings of EuroVAST 2010, international symposium on visual analytics science and technology, Bordeaux, France, 8 June

PoliVisu (2020) Policy development based on advanced geospatial data analytics and visualisation. www.polivisu.eu. Accessed on September 2020

PoliVisu Policy Visuals Toolbox (2020) www.polivisuals.eu

Sanderson I (2002) Evaluation, policy learning and evidence-based policy making. Public Administration, Wiley Online Library

United Nations (2020) E-government survey 2020—Digital government in the decade of action for sustainable development, United Nations Department of Economic and Social Affairs

Yannis Charalabidis is Full Professor of Digital Governance in the Department of Information and Communication Systems Engineering, at University of the Aegean. In parallel, he serves as Director of the Innovation and Entrepreneurship Unit of the University, designing and managing youth entrepreneurship activities, and Head of the Digital Governance Research Centre, coordinating policy making, research and pilot application projects for governments and enterprises worldwide. He has more than 20 years of experience in designing, implementing, managing and applying complex information systems, in Greece and Europe. He has been employed for eight years as an executive director in SingularLogic Group, leading software development and company expansion in Greece, Eastern Europe, India and the US. He has published more than 200 papers in international journals and conferences, while actively participating in international standardisation committees and scientific bodies. In 2016 he was nominated as the 8th most productive writer in the world, among 9500 scholars in the Digital Government domain, according to the Washington University survey. He is 3-times Best Paper Award winner in the International IFIP e-Government Conference (2008, 2012, 2016), winner of the first prize in OMG /Business Process Modelling contest (2009) and 2nd prize winner in the European eGovernment Awards (2009). In 2018, Yannis was nominated among the 100 most influential people in Digital Government worldwide, according to the apolitica.co list.

Chapter 6
Turning Data into Actionable Policy Insights

Jonas Verstraete, Freya Acar, Grazia Concilio, and Paola Pucci

Abstract It is becoming clearer that data-supported input is essential in the policy making process. But at which point of the process, and in which format, can data aid policy making? And what does an organisation need to turn data into relevant insights? This paper explores the role of data from two perspectives. In the first part, data and data analysis are situated in the policy making process by mapping them onto the data supported policy making model and highlighting the different roles they can assume in each stage and step of the process. The second part discusses a practical framework for policy-oriented data activities, zooming in on the data-specific actions and the actors performing them in each data-supported step of the policy making process. We observe that a close collaboration between the policy maker and data scientist in the framework of an iterative approach permits to transform the policy question into a suited data analysis question and deliver relevant insights with the flexibility desired by decision makers. In conclusion, for data to be turned into actionable policy insights it is vital to set up structures that ensure the presence and the collaboration of policy-oriented and data-oriented competences.

Keywords Role of data · Data for dialogues · Making policies with data · PoliVisu project

J. Verstraete · F. Acar
Dienst Data En Informatie, Bedrijfsvoering, Stad Gent, Belgium
e-mail: Jonas.Verstraete@stad.gent

F. Acar
e-mail: Freya.Acar@stad.gent

G. Concilio · P. Pucci (✉)
Department of Architecture and Urban Studies, Politecnico di Milano, Milan, Italy
e-mail: paola.pucci@polimi.it

G. Concilio
e-mail: grazia.concilio@polimi.it

© The Author(s) 2021
G. Concilio et al. (eds.), *The Data Shake*,
PoliMI SpringerBriefs,
https://doi.org/10.1007/978-3-030-63693-7_6

6.1   Introduction

The manifold experimentations conducted in the last years in the use of data—big, open data—have shown the great potential of these sources for addressing the real time monitoring of urban processes and operational actions (i.e. solving crisis situations such as traffic jams and accidents, or schedule adjustments in transport supply). However, their use and impact within the policy making processes is still a more controversial and less obvious question to be addressed.

Critical ex-post evaluations on the potential and limits of data-informed policy making have also led several public bodies such as the EU and the US Congress to understand and deepen the possibilities and the challenges related to the use of data in policy making and analysis in specific areas of application (De Gennaro et al. 2016; Jarmin and O'Hara 2016; Lim et al. 2018).

Underlining the relevance of data in promoting dynamic resource management; in allowing the possibility to discover trends and to analyse their developing explanation; in fostering public engagement and civic participation and, finally, in sustaining the development of *"robust approaches for urban planning, service delivery, policy evaluation, and reform and also for the infrastructure and urban design decisions"* (Thakuriah et al. 2017, p. 23), become central issues for cities that already have lots of data and, thanks to fast-evolving technologies, see the growing opportunity to collect more and faster.

Thanks to open data initiatives and emerging data ecosystems, data is shared across a multitude of actors. Policy makers, however, do not need data, they need insights. Data visualisations and advanced analytics can provide these insights, but only if they give the right answer to the right policy question. How can data experts assure this match? How and where do data activities fit into a policy making process? And what are the key aspects for an organisation to turn data into gold?

By dealing with these questions, the PoliVisu project aims to guide public administrations in adopting a data supported policy making process, developing a theoretical model describing the different stages of policy making and the role data can play in each step of the process. This topic is discussed in the first part of the paper, where practical examples drawn from the PoliVisu pilot's direct experience are shared to further the understanding of the model and the possible uses of data for urban policies. The second part focuses on the different typologies and ways of implementing relevant data-related activities that should be carried out in a policy making context to define and answer a policy question effectively.

6.2   Policy Making Supported by Data

In PoliVisu, we assume to work with a model based on the policy cycle that means conceiving policy as a process, by conceptualising it as a data-assisted policy experimentation cycle consisting of interrelated, stepwise or cyclical stages.

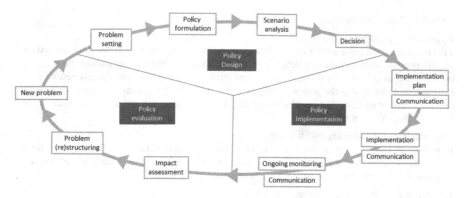

Fig. 6.1 The policy cycle model

The type and role of data analysis can change at each stage of the policy making process.

The policy cycle model (Fig. 6.1) consists of 3 stages: policy design, policy implementation, and policy evaluation. Every stage consists of several steps. The stages and steps do not follow each other in a linear manner, rather they are defined as overlapping and cyclical. Moreover, the stages of policy making tend to become more integrated and overlapping when data is involved (Concilio and Pucci 2021).

Before discussing this model in more detail, a common understanding is required of the data-based analysis types that can be employed during the policy making process. Here in Table 6.1 we provide the definitions, starting with relatively simple analysis types and proceeding to more complex ones.

All the previous data analysis types and techniques can be employed in a data-supported policy cycle model which typically consists of three stages: design, implementation, and evaluation.

6.2.1 Policy Design

The first stage of the policy cycle model is policy design (Fig. 6.2). The policy design stage is focused on highlighting a collective policy problem, identifying a set of goals and objectives in relation to it and defining policy strategies and actions to contribute to solving the problem. The essential steps of policy design are problem setting, policy formulation and scenario analysis.

Problem setting highlights the existence of a problem and legitimizes it as a collective problem to be faced. It consists of an analysis of the existing policy and how it deals with the problem. It also includes a reconstruction of the public debate, and the identification of the stakeholders and actors potentially involved. In the problem setting step, data can be useful to explore the effectiveness of past policies and to better know the current urban phenomena affecting the problem. The depiction

Table 6.1 Types of data analysis

Exploratory analysis (What is the data?)

This is the first, but most crucial part of data analysis. The purpose of an exploratory analysis is to gain insights in data characteristics, to assess the potential of the data, to answer the policy question and to get ideas for the analysis. An exploratory analysis does not directly result in information to be used in the policy making process. It is merely a preparatory step for the main analysis in order to define the usefulness and quality of a data source and to gather information for designing the main analysis

Reporting and monitoring (What is happening?)

Reporting and monitoring are considered as the most basic type of analytics. Data is cleaned and new features might be created through integration and aggregation of raw data features. These original and derived features are then visualised, often in real time

Descriptive analysis (What did happen?)

A descriptive analysis describes the situation through standard statistical analysis methods. This usually includes averages, general trends, relations, and variations of a variable in several scenarios. Although a descriptive analysis can detect and highlight correlations between observations, it should not draw conclusions on causal relations. Moreover, when using only descriptive statistics (mean, mode, ranges of the variables), the conclusions of the descriptive analysis should stick to what is seen in the data. To deduce properties of a larger real-life population, that is beyond the data sample, inferential statistics must be used in the descriptive analysis. Some descriptive analysis results might be integrated into reporting dashboards or monitoring visualisations

Diagnostic analysis (Why did it happen?)

Knowing what happened is the first step, but it is not enough to make a confident decision. A diagnostic analysis aims at explaining the findings of the descriptive analysis. This often requires a combination and analysis of other data sources. By diving deeper into multiple data sources and looking for patterns, a diagnostic analysis tries to identify and determine causal relationships. More advanced statistical methods such as probability theory and regression analysis can be used to test hypotheses about why something is visible in the data. Also, machine learning techniques can help recognising patterns, detecting anomalies and identifying the most influential variables

Predictive analysis (What will happen?)

Once the diagnostic analysis allows an understanding of why something happened, predictive analysis can help determine what can be expected to happen next. Of course, all predictions have their shortcomings and should be handled with care. Still, having some information on potential future scenarios will help policy makers to make better decisions. Moreover, the continuous development of modern analytical techniques and the availability of big data, will enable more and more organisations to use predictions with fast-increasing reliability. A predictive analysis takes as input a series of independent variables. Statistical models and artificial intelligence techniques are then used to predict the most likely outcome, the dependent variable. Predictive analysis techniques are based on the development or training of a predictive model. Data scientists need to work closely together with policy makers to avoid poor business assumptions and ensure the predictive model makes sense. Also, qualitative training data must be available

(continued)

Table 6.1 (continued)

Prescriptive analysis (How can I make it happen?)
Prescriptive analysis helps to make decisions about what to do to attain a desired outcome. A prescriptive analysis starts from a predictive model and adds constraints and business rules to it. Prescriptive analyses are suited when the number of variables to be taken into account and the amount of data to digest, exceed the human capabilities. Prescriptive analyses require business rules and constraints to be precise. This requires close collaboration of the data scientist with the policy maker or decision maker to ensure the analysis provides meaningful recommendations. Prescriptive models can be very complex, and the appropriate techniques must be applied to consider all possible outcomes and prevent erroneous conclusions. As for predictive analysis, the key to prevent costly mistakes is training and testing the model

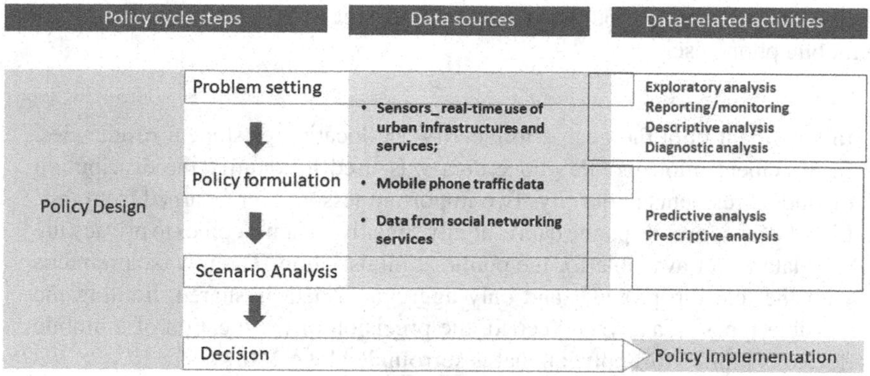

Fig. 6.2 Policy design cycle and data-related activities

of ongoing trends and the consequent definition of the problem can be supported by the collected data. An as-is representation aids understanding the problem at hand.

In this step exploratory, descriptive, and diagnostic analysis methods can help to understand the data, describe the properties of the problem through the available data, and define the exact dimensions of the problem. Since the problem setting step is merely aimed at defining the policy problem and not at finding solutions, the predictive and prescriptive analysis methods are not suited for this step.

Several examples of how data can assist the assessment of a policy problem are given below.

Through traffic sensors and floating cars, we can collect data in real-time related to the movement of vehicles, their speed, and the occupancy of a road. With smart cards for the Public transport users and vehicle sharing systems we can learn the position and information related to each user. In both cases, data can help analysing the functioning of urban infrastructures and services.

As a regional capital, Pilsen suffers from traffic congestion because of the city design, increasing traffic and the organization of mobility and transport. In the Pilsen Pilot a traffic dashboard has been developed that shows congestion in real time, allowing to identify at which location congestion is present, how severe it is and how long it lasts. This aids in identifying the problem and getting information on the severity and the complexity of the policy problem.

Mobile phone traffic data can be used to consider the position of each device connected to the cellular network (and, consequently, of the person who owns the device). From this, we can learn mobility patterns of the owners, time-space variability of population distribution in cities and classification of urban spaces according to mobile phone uses.

In the Ghent pilot the goal is to identify the location of student residencies. In the Ghent pilot, mobile phone data was used to identify the distribution of student residents in the city. Two important lessons were learned from this. First, because mobile phone data is highly sensitive when it comes to privacy the raw data is not available for the public administration. The raw data remains with the telecom provider and only aggregated data is shared, limiting the possible types of analysis. Second, the precision of the location of a mobile device is limited to a polygon that is surrounded by cell towers.

Data from social networking services (such as Facebook, Twitter, Instagram, WeChat and others). From this, we can learn information about the location of city users and about the activities they are participating in, daily travel patterns, opinions, feelings and (self)track of habits, performances, and behaviours.

The Ghent pilot attempted to employ social media data to determine the behavioural patterns of students but had to conclude that the data could not be used. First, there was only a relatively small, and probably biased, number of users. Second, the data provided on e.g. location referred to a general point within the city and was not a reflection of the position of the user.

The Issy-les-Moulineaux pilot offered the opportunity to test some tools to crowdsource data that can evaluate the measures put in place during the local main event. During this event approximately 25.000 people were in town over a few hours. The usefulness of the tools to detect any issues related to transport and mobility through a sentiment analysis with this tool was identified. Unfortunately, data wasn't useful due to limitations to access data of the most used, by citizens, social networks (Facebook and Instagram).

The second step in the policy design stage, policy formulation, is directed towards the identification of shared objectives and the alternative options for intervention in relation to the problem defined in the previous phase.

In this step, predictive and prescriptive analysis methods can be employed to support the choice between alternative measures. For a given policy problem several possible policy measures might exist. In this step the pros and cons of every measure are investigated and eventually one policy measure is chosen. Experimental iterations in the policy making process can be used to diagnose the effect of different scenarios. At the same time, these iterations are a good opportunity to gather training data, validate and refine predictive models.

The pilot of Issy-les-Moulineaux developed a mobility dashboard to visualise and identify the most important congestion points in a detailed way to support the policy makers in taking decisions and defining policies. As a first result, it became clear that the bulk of traffic originates from cars passing through, and not from inhabitants of Issy-les-Moulineaux. In collaboration with a local start-up an application was tested in congested areas to propose different paths, defined by the City on precise data (and not just on algorithms), and to communicate to drivers in real time.

The last step of the policy design stage, scenario analysis, can be carried out once a policy measure has been chosen by the use of different methods to "create a set of the plausible futures" rather than "forecasting of the most probable future" (Amer et al. 2013, p. 25). Such a different focus explains why scenarios are not appropriate in forecasting but rather in "backcasting", that is, identifying desirable futures and the action required to attain them.

By considering existing trends and possible future developments, thanks to predictive and prescriptive analysis of data, it becomes possible to assess the potential benefits and costs of different alternative scenarios and, by doing this, making a decision.

6.2.2 Policy Implementation

The policy implementation stage (Fig. 6.3) is focused on the realisation of a policy plan. In this stage, the monitoring of early impacts is the key. The essential steps are

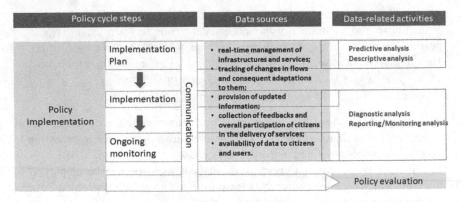

Fig. 6.3 Policy implementation cycle and data-related activities

the making of the implementation plan, its realization (implementation), on-going monitoring, and communication.

The implementation plan is necessary for policy implementation to be as effective as possible. While doing this the opportunities for data-based monitoring activities should be considered. When a data collection plan is included in the implementation plan, data collection infrastructures need to be designed together with data analysis for the implementation step.

> In the Mechelen pilot a regional traffic model is being used to study the traffic in the city. A recent policy decision introduced the concept of "school streets", streets that are being closed at the beginning and at the end of the school day. The traffic model combined with local traffic count data measures and analyses the impact of school streets on traffic behaviour in and around the school streets.

In this step, data can be useful to guarantee the full impact of the policy implementation and the achievement of the policy goals. Descriptive and predictive analysis methods are the most relevant in this step, because data can help to describe the current situation and context, namely all the spatial and socio-economical aspects affected by and involved in the process- the policy decision is directed to. From the other, predictive methods can be designed to foresee the impact of a policy decision and how the decision will affect the context.

The implementation of a policy might require a lot of time and can produce important temporary effects in the context. In this step, data can be useful to verify that policies are implemented as planned and to check early policy impacts. Insights from reporting and monitoring activities and diagnostic analyses can support policy tuning where needed.

This step is crucial because it can be performed in an experimental way that considers the data generated as a result of the impact of the policy interventions so far. In fact, if an effective monitoring plan is associated with the implementation plan, the implementation step can be sided by a step of on-going monitoring. In this step, data can be collected concerning all aspects that are hypothesized to be influenced by the policy decision, just as it is being implemented. An observation of the context on a daily basis can in fact create a rich learning opportunity both for the institutions as for the citizens in the context.

During the entire policy making process communication is essential. This includes communication with citizens, communities, and other stakeholders, allowing them to participate in the policy making process and even take part in the data collection process. To be most effective communication should take place in parallel with the implementation of the policy decision.

6.2.3 Policy Evaluation

The policy evaluation stage (Fig. 6.4) examines the desired and undesired impacts achieved through the implementation of a policy. It monitors how the policy contributed to address the initial problem, whether possible disadvantages were avoided, which advantages arose and how a policy is likely to perform in the future.

How a policy should be evaluated needs be decided when a policy problem is defined. Policy evaluation takes place throughout the whole policy cycle with a final evaluation stage at the end. The definition of the issues to be faced and the objectives to be achieved already determine what data will be relevant for evaluation, and what procedures for data collection and evaluation need to be established. An evaluation not only refers to final results, but rather to the whole planning and implementation progress.

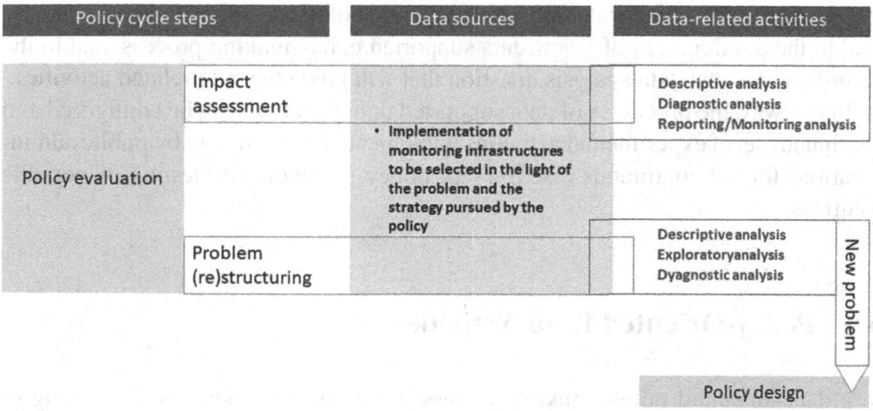

Policy cycle steps		Data sources	Data-related activities
Policy evaluation	Impact assessment		Descriptive analysis Diagnostic analysis Reporting/Monitoring analysis
		• Implementation of monitoring infrastructures to be selected in the light of the problem and the strategy pursued by the policy	
	Problem (re)structuring		Descriptive analysis Exploratory analysis Dyagnostic analysis

New problem

Policy design

Fig. 6.4 Policy evaluation cycle and data-related activities

To be able to perform a policy evaluation, a data collection plan is included in the step of Impact assessment for supporting, at least:

- multidimensional qualitative and quantitative impact assessments;
- observation of direct and/or indirect effects—for instance at the urban level;
- participatory evaluation—i.e. shared with, and possibly affected by, the very same stakeholders involved in the policy implementation process.

During the *impact assessment* step exploratory, descriptive, diagnostic, and prescriptive analysis methods can be useful, next to reporting and monitoring activities.

In the city of Pilsen, the Sustainable Urban Mobility Plan (SUMP) is currently being implemented. It is a mobility plan, up to the year 2025, which includes 82 measures for better mobility in the city. In the PoliVisu project we created tools for visualising the state of traffic before, during and after the implementation of SUMP measures. These visualisations help to evaluate the impacts of the measures.

Thanks to the analysis carried out in the *impact assessment* step, it is possible to discover the policy results and to evaluate how successful the policy was, based on expectations set during the policy design stage. This approach allows us to critically revise the contents of the policy measures, as well to reconsider the nature of the problem itself. It is possible to obtain new insights on the characteristics of the problem, on its evolution over time and on possible new deployable strategic responses to tackle it.

The last step, problem restructuring, represents a moment of retrospective reflection in which the descriptive and diagnostic analysis from the data can contribute to reconsidering the initial problem. A new definition of the problem will consequently lead to the development of a new data supported policy making process, and to the definition of a new data analysis question that will guide the data related activities.

In this way, the processes of data supported policy making can be configured as a continuous set of experimental activities implemented dynamically by public administrations for the continuous discovery of policy problems and testing of possible solutions.

6.3 Policy-Oriented Data Activities

In a data-supported policy making process, most effort is spent on identifying or developing tools that analyse and visualise a well-defined type of data in order to support the decision process. In cases where the required data is already known, and the information question is well advanced, this approach can be effective. However,

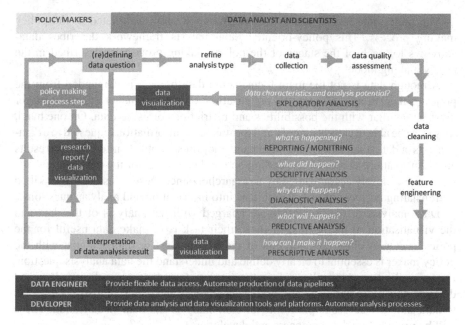

Fig. 6.5 Policy-oriented data activities framework

in the policy making context most policy problems still require a translation to a more specific information question and relevant data is often not easily available or identifiable. Also, data visualisations might need frequent adaptations to a varying audience and to the specific story the policy maker wants to communicate. On top of this flexibility, policy makers often have the urgency to respond to problematic circumstances. The combination of all these requirements asks for a different approach to the data activities supporting the policy making process.

Figure 6.5 provides a practical framework developed with the experiences of the PoliVisu project. The framework in which the policy makers and their support teams operate. It describes a policy-oriented approach to data-analysis and data visualisation. This approach suggests a close collaboration between data literate policy makers and data specialists. The iterative nature of the collaboration aims to ensure the data analysis is well customized to support the policy question.

6.3.1 Differentiating Roles and Competences

The framework considers the different roles, tasks and competencies involved in the process. The main actors are the policy maker, the data analyst or data scientist, the data engineer and the developer. In small organisations, multiple roles can be covered by one single actor, but given the specific competencies associated with each role, this is not recommended for larger organisations and more complex policy questions.

The policy makers main activities and competencies are related to the policy making process. This policy-oriented data-analysis framework describes data-activities for each of the stages of the policy making process as described in the policy making model.

Although data is not the main focus, some data literacy is demanded from the policy maker. In a data-supported organisation, a decision maker or policy maker must be familiar with the possibilities and restrictions of using data. On one hand, they must be able to clearly identify and formulate the information requested to a data-analysts or data scientists. On the other hand, they must be able to interpret data results and visualisations correctly. Depending on the data literacy of the decision makers, they can be closely accompanied by a researcher, since these roles are often skilled in translating business or policy problems into information and analysis questions.

Data analysts or data scientists are charged with the analysis of the data and the visualisation of the results. In short, their task is to make data useful for the policy maker. A close collaboration of the data analyst or data scientist with the policy maker is essential to clearly define and understand the data analysis question. Data visualisations play a key role in the conversations between policy makers and data specialists. To perform their tasks, data analysts and data scientists need access to data and have access to platforms for data-analysis and data visualisation. These will be provided by data engineers and developers.

Data engineers are responsible for providing access to data, the main resource for data analysts and data scientists. To enable quick and flexible responses to policy information questions, data engineers must organise quick and generic access to (raw) data for exploration and analysis. Some of the data will be explored but not used in the final data analyses or visualisations. If the final analysis result is considered useful, the data access and data streams will need to be automated in production data-pipelines to be frequently updated and available for the end-user, the policy maker, via the visualisation. Organisational approaches and methodologies to deal with these challenges, go beyond the scope of this framework. Although the data engineer has a specific role, it is often combined with the other IT-oriented role in the process: the developer. Besides the provision of tools and platforms, the developer can contribute to the automation of analyses.

6.3.2 Balancing Flexibility and Usability

The required velocity and flexibility of policy-oriented data analyses do not allow a conventional development lead time for every new data visualisation. Similar to data engineers, developers need to provide generic and flexible platforms for analysis and visualisation. Two kinds of tools can be distinguished based on the target users and the trade-off between flexibility and versatility on one side, and usability and required expertise on the other side.

From one side, data analysts and data scientists need flexible and versatile analysis platforms. Since they are considered data experts the interfaces can be built for advanced users and might even be code-based. These platforms include a wide range of analysis and visualisation methods, but graphical customization options are often limited. Since data scientists are expected to deal with large and big datasets, computation performance is an important requirement for these analysis platforms.

From the other, the second type of tools required for policy-oriented data analysis still provides a wide range of visualisation options, allows some data manipulation, and deals with multiple data types. It combines this flexibility with easy to use interfaces and visual customization options. Data scientists and data analysts can use these tools to communicate analysis results with policy makers via dynamic visualisations. Less technically skilled data analysts and policy makers themselves will use these generic tools to explore data and analysis results, discover insights and build basic dynamic visualisations. These self-service tools provide data analysts and policy makers with large independence to quickly respond to policy information questions without the need for development skills or budget and time consuming IT-projects.

6.3.3 Transforming Iterations into Experimental Drivers

The policy-oriented data analysis framework shows common analysis steps and analysis types from a data analyst and data scientist viewpoint. The framework proposes an iterative approach to enhance flexibility and agile refinement of the analysis to fit the policy questions.

Each analysis iteration starts with defining or refining the data analysis question and the appropriate analysis type and method that will be used in the next iteration. The next step is the identification and collection of input data. It is highly recommended that a quality assessment is conducted on the collected data to ensure or estimate the reliability of the results that will be used for policy making.

> In the pilot of Ghent the goal is to identify the location of student residencies in order to assess the pressure on the housing market. In the city administration several information sources on residential students were used in different departments. These datasets were gathered, cleaned and their quality was evaluated. The quality assessment revealed that the quality of the data was not sufficient and that the necessary information was not present to measure the real impact.

When starting with a new data analysis question or new data, it is recommended to perform an exploratory analysis to 'get to know' the data. The purpose of the exploratory analysis is to gain insight into data characteristics, to assess the potential of the data to answer the policy information question and to get ideas for the main analysis. Data analysts or data scientists explore the data using basic visualisations and summary statistics. This exploration should not take too long but it is crucial for the efficiency of the next analysis steps. Since multiple data sources might be explored, this iteration can be repeated multiple times for one policy information question.

In the Pilsen pilot, a traffic model is constructed to assess and monitor traffic flows thanks to a combination of modelling data and live traffic sensor data. A lot of possible data sources can be entered into the model, such as geo-time series of detectors and their interpolation, Police (Municipal/ČR) events and car accidents. These datasets have been explored and samples of the datasets have been used for preliminary visualisations.

Before starting the actual data analysis, the collected data often needs some manipulation. Data cleaning is about detecting and correcting unexpected, incorrect, inconsistent, or irrelevant data. The quality of the input data will determine the usability of every analysis result, no matter how complex the algorithm that is used: qualitative data always beats fancy algorithms. Data scientists spend a considerable part - up to 60% - of their time on cleaning and formatting data. It is clear that access to clean and structured data can save a lot of time and proper cleaning is essential for reliable results to support decision making.

Feature engineering is the process of creating new input features by combining or categorising the existing features of the raw data. Which features will be created depends on the analysis the data will be used for. This step requires a good understanding of the data definitions and involves domain expertise. This means assumptions made by the data analyst or data scientist in this phase should be carefully discussed with the policy makers.

After completing the data cleaning and feature engineering steps, the raw data has been transformed into an analytical base table.

In the Ghent pilot, data related to the policy problem was not available from administrative data sources. Therefore, new possible data sources were explored, such as telecom data and Wi-Fi sniffing data. To effectively understand the data that would be delivered by the telecom provider several meetings were organised with the data scientists from the company. This allowed the company to fully understand the analysis question, to construct a plan together and for the members of the local administration to understand the data results they would be receiving.

In Flanders, the Federal police road accident data has been used to map road accidents to specific locations on a map. PoliVisu was able to map 87% of the road accidents consistently on a map for the last five years. Several meetings with specialists from the police and the traffic safety institutes lead to an interactive map, including the location of schools. New data sources combined with interactive co-creative session commences furthering data manipulation experiments with ANPR data that gives more insights in traffic intensity and driving speed. The final result will be a traffic safety map instead of a road accident map.

The analytical base table can now be used in analysis algorithms to derive new information from the data. The different types of analyses are discussed in paragraph 6.2. At the end of the analysis iteration, the results are visualised and interpreted.

The iterative approach in this model suggests not to get stuck on the design of the perfect data analysis question. Instead, a first analysis iteration is executed with the basic analysis ideas. Sometimes the first results can already serve the policy maker or can be used to redirect the analysis question. Gradually, more complexity can be added to the analysis, learning from the previous iterations. The increasing complexity can be related to the policy question and the amount of variables that need to be taken into account. Analysis iterations might also gradually add complexity to the analysis method that is used. It is common to start with descriptive analysis, evolving to diagnostic and predictive analysis, to finally develop prescriptive analysis models to drive decisions.

The goal of the pilot of Issy-les-Moulineaux was to achieve a shift in behaviour concerning car use through a communication campaign. In the first, simpler, analysis iteration it was shown that only 27% of the local population uses the car, and that the congestion problem is mainly caused by traffic passing through. This allowed us to adapt the implementation plan and perform more specific and complex analyses in the second iteration.

6.4 Conclusions

Although data seems to be everywhere nowadays, finding suitable qualitative data is often the first obstacle to be overcome in data supported policy making. Turning the data into relevant insights is the next big challenge. Data visualisations and analyses can provide these insights if the policy question is well defined and correctly interpreted. Different data analysis types will be used depending on the phase in the policy making process. Performing the data analysis and the creation of data visualisations in an iterative way, enables the data analysis to be adapted to the needs of the policy maker while gradually increasing the complexity. The introduction of an explicit collaboration between the data literate policy maker and the data experts during these iterations, will ensure that the data response properly fits the policy question. Data visualisations with intermediate results will support this collaboration. The use of flexible data platforms and generic tools for data access, analysis and visualisation can provide the versatility and velocity requested by policy makers. The pilot experiences in the PoliVisu project permitted the development of a data supported policy making model and a practical framework for policy-oriented data activities.

References

Amer M, Daim TU, Jetter A (2013) A review of scenario planning. Futures 46:23–40. http://dx.doi.org/10.1016/j.futures.2012.10.003

Concilio G, Pucci P (2021) The data shake: an opportunity for experiment driven policy making. In G Concilio, P Pucci, L Raes, G Mareels (eds) The data shake. Opportunities and obstacles for urban policy making. Springer, PolimiSpringerBrief

De Gennaro M, Paffumi E, Martini G (2016) Big data for supporting low carbon road transport policies in Europe: applications, challenges and opportunities. Big data research 6:11.25

Jarmin RS, O'Hara AB (2016) Big data and the transformation of public policy analysis. J Anal Manag 35(3):715–721

Lim C, Kim KJ, Maglio PP (2018) Smart cities with big data: reference models, challenges and considerations. Cities 82:86–99

Thakuriah P, Tilahun NY, Zellner M (2017) Big data and urban informatics: innovations and challenges to urban planning and knowledge discovery. In: P. Thakuriah et al. (eds) Seeing cities through big data. Springer Geography

Jonas Verstraete is a spatial data and information management expert. He has a background as Msc. in biosciences and 10 years experience in data management and analysis. Jonas currently works at the Data and Information Office of the City of Ghent on information management and the preparation and analysis of data for decision making. In the last 2 years Jonas has worked extensively on the organisational and strategic aspects of a data driven organisation.

Freya Acar Project manager for European and Flemish projects concerning (open) data, data driven policy making and smart city for the city of Ghent (Belgium). She obtained her MSc in Theoretical and Experimental Psychology from the University of Ghent in 2014. Hereafter she started a PhD at the Department of Data-Analysis concerning the assessment and correction of bias in neuroimaging studies. Communicating data questions and results to peers with less data affinity was one of the key aspects of her PhD, which is further explored through visualizations in the PoliVisu project.

Grazia Concilio Associate professor in Urban Planning and Design at DAStU, Politecnico di Milano. She is an engineer and PhD in "Economic evaluation for Sustainability" from the University of Naples Federico II. She carried out research activities at the RWTH in Aachen, Germany (1995), at IIASA in Laxenburg, Austria (1998) and at the Concordia University of Montreal,

Canada, (2002); she is reviewer for several international journals and member (in charge of LL new applications) of ENoLL (European Network of open Living Lab). Team member in several research projects; responsible for a CNR research program (2001) and coordinator of a project funded by the Puglia Regional Operative Programme (2007–2008) and aiming at developing an e-governance platform for the management of Natural Parks. She has been responsible on the behalf POLIMI of the projects Peripheria (FP7), MyNeighbourhoodlMyCity (FP7), Open4Citizens (Horizon 2020 www.open4citizens.eu); she is currently responsible for the Polimi team for the projects Designscapes (Horizon 2020 www.designscapes.eu), Polivisu (Horizon 2020 www.polivisu.eu) together with Paola Pucci, and MESOC (Horizon 2020 www.mesoc-project.eu). She is coordinating the EASYRIGHTS project (Horizon 2020 www.easyrigths.eu). She is the author of several national and international publications.

Paola Pucci Full Professor in Urban planning at the Politecnico di Milano, and former Research Director of the Urban Planning Design and Policy PhD course at Politecnico di Milano. From 2010 to 2011 she taught at the Institut d'Urbanisme in Grenoble Université Pierre Méndes France at Bachelor, Master and PhD levels and currently visiting professor at European universities. She has taken part, also with roles of team coordinator, in national and international research projects funded on the basis of a competitive call, dealing with the following research topics: Mobility policy and transport planning, mobile phone data and territorial transformations and including EU ERA-NET project "EX-TRA – EXperimenting with city streets to TRAnsform urban mobility"; H2020—SC6-CO-CREATION-2016–2017 "Policy Development based on Advanced Geospatial Data Analytics and Visualisation", EU Espon Project, PUCA (Plan, urbanisme, architecture) and PREDIT projects financed by the Ministère de l'Ecologie, du Développement et de l'Aménagement durable (France). She has supervised and refereed different graduate, postgraduate and PhD theses at Politecnico di Milano, Université Paris Est Val de Marne, Ecole Superieure d'Architecture de Marseille, Université de Tours. She has been Member of the evaluation panel for the Netherlands Organisation for Scientific Research (NWO, 2017), and Member of the NEFD Policy Demonstrators commissioning panel for the ESRC—Economic and Social Research Council. Shaping Society (Uk), on the topic "New and Emerging Forms of Data—Policy Demonstrator Projects (2017).

Chapter 7
Data-Related Ecosystems in Policy Making: The PoliVisu Contexts

Giovanni Lanza

Abstract The article explores the complexity of the ecosystems that develop around data supported policy making. This complexity, which can be traced back to the multiplicity of actors involved, the roles they assume in the different steps of the decision making process, and the nature of the relationships they establish, takes on new connotations following the rising use of data for public policies. In fact, issues related to data ownership and the ability to collect, manage, and translate data into useful information for policy makers require the involvement of several actors, generating ecosystems where co-creation strategies are confronted with the limits of action of the public administrations within broader social and decisional networks. Based on this background, the article aims to provide, through the analysis of the direct experiences conducted by the pilot cities involved in the PoliVisu project, an overview of the opportunities and challenges related to the impact of data in the evolution of decision making networks and ecosystems in the data shake era.

Keywords Actor-network theory · Data supported policy making · Digital transformation · Decision making ecosystems

7.1 Introduction

Thanks to the development and the widespread diffusion of information and communication technologies (ICT), contemporary cities produce an increasing amount of digital data. This unprecedented availability of information about the behaviours, choices, needs, and desires of large samples of individuals collected on extended time frames allows the adoption of a holistic perspective on emerging urban dynamics in a way that can deeply innovate how public policies are constructed (Rabari and Storper 2015; Kitchin 2014a, b). The literature focuses on this opportunity, underlining the possibility, introduced by the constant availability of digital data, to conceive the policy making process as a continuous experimental activity. In this framework, the

G. Lanza (✉)
Department of Architecture and Urban Studies, Politecnico di Milano, Milan, Italy
e-mail: giovanni.lanza@polimi.it

© The Author(s) 2021
G. Concilio et al. (eds.), *The Data Shake*,
PoliMI SpringerBriefs,
https://doi.org/10.1007/978-3-030-63693-7_7

approach to a public issue is based on the diagnosis of current trends as a starting point to design future scenarios. Then, the policy response is dynamically implemented, and its impact is evaluated by considering, through the analysis of data-based feedback, the reactions and effects produced by the policy itself. Thanks to the ongoing data revolution, decision makers can thus not only rely on an increasing analytical capacity which enables them to shed light on the complex dynamics which involve contemporary cities and their populations, but also to respond more effectively to emerging issues through the implementation of targeted public policies.

However, it is essential to point out that the opportunity to promote institutionalized data-supported policy making processes entails several difficulties whose magnitude is directly proportional to the complexity of the interactive network of public and private actors involved and, clearly, to the complexity of the problems to be addressed.

Among the possible challenges facing public administrations, two seem particularly relevant to this article.

The first challenge concerns the possibility (or the willingness) of a public actor to experiment with the use of digital data to drive its decision making activities (Thakuriah et al. 2017). This opportunity is influenced both by the actor's propensity to innovate and by the position the actor occupies in the institutional framework of reference: a variation in the availability of political and economic resources, that are both necessary in the field of digital innovation, can generate potential imbalances between administrations.

A second challenge concerns the ability to structure a culture of *datacy*, which, in Batini's words (2018), is a measure of the decision makers' capability to collect different data, evaluate its quality, interrogate, and use it to analyse reality and solve complex problems. This perspective is based on the ability (and the necessity) to build political, technical, and legal frameworks to let individuals and government entities to be able, for each phase of the policy making process, to collect, access, produce, and analyse data-based information of which the production and ownership are increasingly fragmented. Thus, cities face the growing need to involve additional actors and expertise, from data owners to technical partners, ultimately increasing the level of institutional complexity of the data-related ecosystems.

The two challenges underline the importance and the need for a behavioural change on the part of decision makers who, due to the effects of the data shake, find themselves involved in increasingly complex and ramified decision making networks. Within these ecosystems, whose level of complexity is linked to the content and relevance of the policy in question, the opportunities and challenges introduced by the data bring the participating actors, some of them new but with increasingly central roles, to interact with each other in new and dynamic ways.

Therefore, public administrations are required to develop the ability to build and manage these ecosystems by limiting potential conflicts between stakeholders and adopting strategies to deal with some critical aspects related to digital innovation, such as the problems of data ownership and the dependency on providers of data and technologies.

In this article, some characteristics of these ecosystems are analysed, with particular attention to the new types of participant actors and to the new data-oriented
relations they establish, referring to the experiences of the pilot cities of the PoliVisu
project that have been directly confronted with this relevant effect of the data shake.

7.2   The PoliVisu Project as a Testbed for Digital Innovation

The PoliVisu project, developed within the European Horizon 2020 program framework, offers an interesting case study to analyse and understand the opportunities
and challenges of digital innovation in urban governance. PoliVisu, in particular,
aims to test the usefulness of data visualisations, collected from different sources,
and related to multiple urban issues, in supporting public sector's decision making
activities. To achieve this aim, in the awareness of the technical-political implications associated with the nature of these experiments, the project promotes the
establishment of a system of relationships and mutual support by including actors
with different expertise and competences in the PoliVisu consortium.

In this macro-network, the pilot cities' administrations have played the most relevant role for the topic of the present article, mediating between the general objective
of the project and the difficulties inherent in dealing with the contextual and concrete
promotion of data supported policy making they have experimented. To face the
challenges they encountered while implementing their activities, partners have built
local micro-networks, involving both other members of the consortium and external
local actors to respond, with a data and visualisation-based approach, to the main
policy questions arising in their specific contexts.

Thanks to the analysis and mapping of the activities of the pilot cities it has been
possible to paint a policy network canvas for each municipality, whose characteristics
are discussed in the following sections of the article.

This analysis has allowed us to capture to what extent each city took the opportunity to use the resources and the context of PoliVisu to test a new and different
approach to policy making and governance. The identified ecosystems, which are
flexible and evolving networks of relationships, exist and change according to the
emerging public issues to be faced. In fact, the contents of a policy affect the type
of actors to be involved in the network and the choice and availability of data to
collect, analyse, and visualise. This condition has emerged in the PoliVisu project,
where the pilot cities differ in size and in their position in their administrative framework of reference. Moreover, they have been confronted with emerging local issues
that, although primarily related to mobility, have been tackled differently due to the
different availability of data and the influence of the local and supra-local political
agendas, thus leading to the construction of networks of varying complexity.

Five pilots are participating in the PoliVisu project. The case of Issy-les-
Moulineaux (France) concerns a medium-small city of the Paris metropolitan region
affected by high levels of car congestion. This condition is mainly due to the city's

economic vitality and commuting transit to and from Paris. Therefore, Issy-les-Moulineaux wishes to communicate real-time traffic information to citizens and develop a control dashboard to support the public administration's operative sectors in their planning activities. The Issy pilot's primary goals are related to establishing an effective and clear communication with the citizens (to achieve a long-term effect of behavioural change in mobility habits) and to test how data visualisations can support decision making processes and facilitate collaboration within the administration.

A different situation characterizes the pilot of Ghent (Belgium), which aims to identify the hidden population of students. Every year many unregistered students reside in Ghent, widely influencing the housing market and the mobility system. Identifying where students live can help the policy making process and produce positive and relevant impacts on the city's liveability.

Also, the pilot of Pilsen (Czech Republic) is mainly focused on mobility. The layout of the city causes several congestion challenges and problems. To monitor and predict the impact of changes, such as roadworks, the traffic model developed for the city will be improved during the project using traffic sensor data to identify and visualise traffic volumes and their changes over time.

The Mechelen pilot (Belgium) has two main objectives. The first objective is to provide traffic modelling for the city. Mechelen, together with the police zone Mechelen-Willebroek, has integrated an existing traffic model and aims to enrich its usefulness by adding data from ANPR (Automatic number-plate Recognition) cameras. The traffic model will be used as a predictive tool to monitor the impact of planned road works in the city. The second objective deals with the recently introduced "school streets" where traffic calming solutions such as road closures are in place. The aim is to measure and analyze traffic and congestion variation (both in the closed streets and the neighboring streets) before and after implementing the measure.

Finally, the Flanders pilot (Belgium) concerns the traffic accident map, a tool that visualises the traffic accident data which is obtained from the federal police, allowing specialized analytics directly on map (through heatmaps and application of advanced filters).

It can be noted that the pilot cities are very different both in terms of the issues they deal with and in the roles played by the involved actors within the decisional ecosystems. However, the analysis of the activities and networks built by the pilots, which is presented in the following sections of this article, shows that evident similarities can be identified in the challenges they face and the strategies they are putting in place to tackle them.

The mapping of the activities and networks that have developed thus becomes a valuable resource both during the implementation of the experiments and in a subsequent phase of process evaluation to manage and understand the complex decision making ecosystems of the digital age.

7.3 Actors and Roles in Data-Related Policy Making Ecosystems

In the field of public policy analysis, the social network that develops around any policy decision represents a fundamental object of study. In fact, the reconstruction of the stages of a decision making process allows us to understand how the participant actors, the relationships they establish, and the resources they spend, have contributed to achieving a certain outcome, namely, the public decision that is the object of analysis (Dente 2011). Although this analytical technique seems more suitable to study processes in a posteriori perspective, therefore considering decisions that have already been taken, it is still interesting to try to apply it in the context of PoliVisu and the pilots' activities, where the experimental nature of the project introduces an additional element of complexity. In fact, the challenges faced by pilots in their path have led to a deeper reflection on their internal organization and the possible evolution of their activities and role in the era of digital innovation.

As a result, it seems reductive to analyse the cases starting from the assumption that the pilots have linearly shaped their activities by focusing only on the initial problem throughout the process. Rather, starting from a specific issue that would have represented an interesting field of data and visualisation experiment, the pilots have generally organized their own micro-network and changed their behaviour to respond to the difficulties they encountered in progressing their activities.

For this reason, the analysis of the pilot cities' networks is related to processes that are still ongoing where the local public administrations keep on experimenting with new kinds of interactions and opportunities. In this perspective, the experimentation becomes a useful approach to guide the public sector in the transition to the digital age acquiring, in the process, a new awareness of its role and responsibilities.

The analysis carried out for the PoliVisu cases is mainly based on the model proposed by Dente (2011) but introducing some changes to adapt the model to the innovations brought by the data shake on decision making ecosystems. Among these, one of the most important appears to be the widening of the range of actors involved. Establishing an agile and effective data supported policy making is an interdisciplinary challenge that requires the co-presence of multiple perspectives. These different viewpoints are related to specific competencies that are not easily found in one multi-talented person. PoliVisu's experience confirms that, in the digital age, the presence and collaboration of different expertise and the availability of multiple viewpoints becomes an essential resource (Wiliford and Henry 2012) to create an environment in which decisions and the use of data to support them can be developed in an integrated and effective way. Through participant observation and interviews, the actors that each pilot has involved have been identified according to their roles within an actor-network scheme and organized into four categories, which should not be considered impermeable; each actor, in fact, can play different roles in the process, thus contributing to making these ecosystems even more complex (Fig. 7.1).

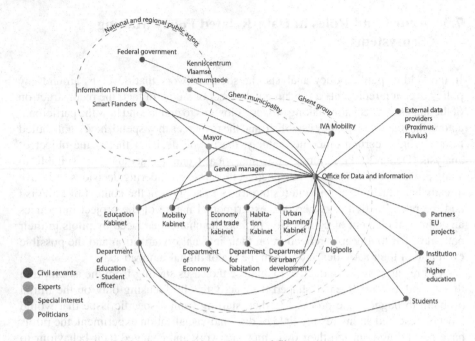

Fig. 7.1 Example of the actor-network from the case of Ghent

Politicians justify their actions based on available political resources given by their status of elected representatives of citizens. In the PoliVisu pilot's case, each context shows different configurations of the participation of politicians in the decision making network. This outcome is due to the fact that the cities involved represent different countries, each one characterized by its specific structure and culture of governance and policy objectives that may or may not facilitate the promotion of digital innovation paths at the municipal level. Although the pilot cases are very different from each other, the analysis shows that regardless of the level of autonomy of each city in promoting its experimental approach to data supported policy making, the role of supralocal policy actors is relevant since it can enable and facilitate, through the definition of regulatory frameworks, the experimental data-oriented activities of local governments.

Many of the political actors identified in the networks represent complex and diverse organizations. It happens that, for example, the same organization can take on both a political and administrative role. Therefore, all those actors who actively participate in the functioning of the administrative machine and whose action is bound by political guidelines and legal frameworks have been indicated as civil servants. In PoliVisu and other data-related ecosystems, a central role is assumed by the offices for data and information (smart city department). Their task is to facilitate digital innovation by continually mediating between the inputs of political actors, the needs of other public administration members, and the proposals and requests of external partners, such as data providers and technical experts. The civil servants of a smart city office are essentially dynamic figures working in a multidisciplinary environment, in line with the evolution of governance in the era of digital innovation. Their role requires a specific ability to interact with other public administration sectors, responding to their requests and facilitating their activity through a continuous ability to communicate with external partners that provide data and services that the public administration does not own.

These interactions are supported by the third type of actor identified during the analysis, the experts, who possess specific skills or knowledge resources that can be spent in policy making processes. Since public administrations are transforming into data-driven organizations, many of the expertise needed to extract value from data may be available within the structure, and it is for this reason that in the decisional ecosystems of the PoliVisu pilots central actors such as the smart city offices take on at the same time the role of civil servants and experts. However, as shown by the PoliVisu cases, referring to external actors is necessary when this expertise is not available within the organization or when it comes to access to data not directly produced or owned by the public sector.

Indeed, if in the past we could argue that the public sector was holding a large number of datasets to be opened, today we have a growing number of high-value datasets held by private bodies and not being used for a public will. To obtain this data, public authorities need to negotiate and buy it from private companies. External data providers are thus becoming an increasingly important player within decision making ecosystems.

They fall within the fourth and final category, the one of the actors with special interests. The actors that play this role participate directly or indirectly in the process for two main reasons.

The first is linked to the fact that the policy decision in question may affect their interests. This mechanism may lead either to the actor's direct participation in the decisional arena or by influencing (both directly and indirectly) the other actors involved in the construction of the policy. The latter option happens, for instance, with citizens or other policy recipients/targets that may express their favour or opposition to a policy through public opinion and voice.

The second is related to the fact that actors representing special interests can build, by participating into the policy making arena, further opportunities beneficial for their interests, as happens in the case of data and service providers interested in collaborating with public administrations to promote their business, build know-how and gain visibility. The relationship with these actors, which are generally external to the organization but fundamental to access data, requires both the existence and adoption of procurement frameworks and good negotiation skills from the public sector.

The actors and roles that have been identified are the results of a generalization based on the PoliVisu pilots' experience. Observing how cities have extended through their experimental activities these ecosystems even beyond the borders and the actors of the project, it is thus demonstrated that digital innovation challenges are faced by involving several and different actors and expertise. However, it should be stressed that an expansion of the network can, in turn, introduce new challenges that cities must face. In fact, too complex networks featuring a wide variety of points of view expressed by different actors, or very dense ones characterized by a large number of relationships involving at the same time several actors, tend to be less manageable and limit the chances of a successful decision making process. Pursuing a balance between the expansion of the network and its sustainability is one of the great challenges introduced by the data shake that is probably engaging the pilots of PoliVisu and many other cities in the world.

7.4 Data-Related Relations

Digital innovation in policy making requires the presence and joint action of multiple actors and expertise, leading to the consequent expansion of decision making ecosystems. However, the increasing complexity of these processes is due to the growing number of players involved and the nature of the interactions they establish. In this context, the involvement of new actors is functional in developing new and profitable relationships necessary to overcome problems and challenges induced by data and visualisation use in policy making processes. By operating an abstraction, decision making ecosystems can be seen as networks of nodes and lines. Each node represents an actor who, as we have seen previously, takes on one or more roles depending on the interest or duty that drives its participation in the process. Moreover, according to the role it assumes, each actor possesses certain resources that can be economic, political, cognitive, legal, and relationship based. Therefore, the participation and behaviour of each actor within the network will be influenced by the availability of resources and whether the same actor is willing to spend them to achieve its objectives. Consequently, the interaction between actors can essentially be interpreted as an exchange of resources between participants in the same decision making process. The lines connecting the different nodes of the network become, ideally, conduits through which these resources can flow, setting in motion the whole interactive mechanism that gives life to the ecosystem.

In the era of the data shake, networks extend and become more complex because the challenges related to data collection, management, and analysis require the availability of more resources and the involvement of multiple actors who can share them. For this reason, analysing the interactions between the participants of a decision making process and their evolution in the digital age allows us to reconstruct the participation strategies of the actors and understand how they face the challenges encountered in the innovation path.

From this point of view, the PoliVisu pilots represent interesting and concrete cases to analyse these interactive mechanisms that were mapped in the actor-network grouping the types of interactions that occur in six categories (Fig. 7.2).

However, it is worth noting that this categorization introduces a simplification since data-related activities in the network can lead the same actor to interact with each other for different reasons and, consequently, exchange different resources according to the multiple roles it can play. Furthermore, the six categories describe

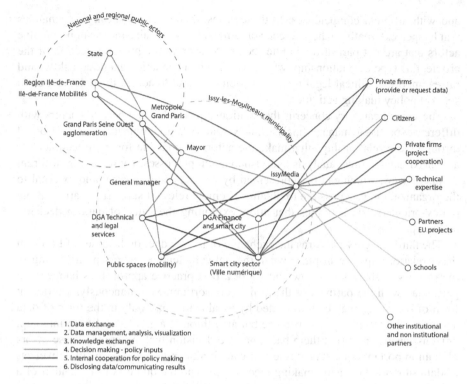

Fig. 7.2 Example of the relations occurring in the actor-network from the case of Issy-les-Moulineaux

interactive processes taking place within most of the pilot cases, underlining how similar kinds of relationships can be established in different contexts.

The first category of relationships, data exchange, refers to the transfer of data between different actors that, in the cases of PoliVisu, generally takes place internally or externally to the public administration that is centrally involved in the experimentation.

In the case of internal exchanges, it is possible that an actor requests data that are produced and owned by other sectors of the same organization. The quality of these interactions strongly depends on the level of *datacy* of the organization itself, as it affects the efficiency of the internal functioning and the regulatory frameworks that distribute the competencies in data production and sharing. For example, it can be observed that these exchanges can be problematic in the typically siloed structures of municipalities where the different sectors are often not well connected between them. In this sense, the data produced or stored by one of the silos is considered a sort of exclusive property and is not shared, even if that may benefit the municipality. Siloed organizational status is reproduced in data production and management.

Such limits can be encountered both within the same organization and in the relations between public organizations belonging to different administrative levels and with different competences. In these cases, therefore, we speak of exchanges that happen externally to the reference organization, which also increasingly involve actors that are not part of the public sector. As seen in the previous section of the article, this type of relationship, which requires considerable negotiation skills and raises important ethical-legal issues, is often essential to access the data needed to support policy making activities.

The second category concerns the relationships established between actors with different expertise to manage, analyse, and visualise data. Certain tasks are distributed through these relationships that make the actors responsible for more technical or advanced data storage, analysis and visualisation processes. By involving different points of view, which can be represented by subjects both internal and external to the organization, it is possible to obtain extremely relevant support to carry out the policy-oriented data activities that allow exploiting the value of data in decision making processes.

The third category concerns the exchange of knowledge. In the cases of PoliVisu these relationships are in place when the network features actors participating in the process to share know-how and adopt a best practice approach as happens, in particular, with the partners of the project consortium. Simultaneously, a different form of knowledge can be transmitted by local actors, not only in the form of data about issues within their competence but also through a greater ability to intercept citizens' needs and bring them back into the decision making arena. These actors play an important consultancy role that can significantly increase the effectiveness of data supported decision making mechanisms and the quality of the content of a policy.

The fourth category introduces another central issue of policy making and concerns the impact of political guidelines on decision making processes. The label policy inputs - decision making identifies the power relations that typically emerge between political actors and civil servants. In the case of policy inputs, the former, using resources linked to the role attributed to them by popular consensus, have the power to establish the political agenda, pointing specific issues of public interest and orienting the consequent policy responses. They also have the power to establish legal frameworks to regulate the use of data for innovation policies, thus guiding the activities of the operative sector of the public administration.

On the other hand, this relationship may also concern the transmission of data and visualisations in the opposite way, from the operational sectors of public administration to the policy makers. This opportunity is extremely relevant since it can become a key strategy to promote and enhance the efficiency of decision making activities.

Important, in this sense, is also the possibility to use data and visualisations to facilitate interactive processes between members of different operational sectors within the same organization. The fifth category identifies these relationships as internal cooperation for policy making, which includes all relationships and exchanges of resources intended to facilitate the sharing of data, information, and knowledge between different sectors, increasing their permeability and capacity for interaction. Although each operational sector of public administrations is generally responsible for some public issues according to a siloed logic, promoting modes of organic

action and cooperation can be a way to make the organization's activity more efficient, thus avoiding the waste of valuable resources and promoting, at the same time,
a *datacy*-oriented approach.

Finally, a last category of relations can be established between the public administration and the recipients of the policy decision, through disclosing data and communicating results. Especially in the case of Issy-les-Moulineaux, the digital revolution
taking place in the municipality is interpreted as a valuable opportunity to strengthen
the link with citizens providing, under the form of open data, both relevant information that is useful to organize their everyday life and insights about the activities of
the municipal administration. Moreover, this type of relationship can be configured,
depending on the intentions and the regulatory framework of each context, as the
result of a strategy of direct involvement of citizens in the decision making arena,
legitimizing their participation as representatives of special interests affected by the
policy decision. However, engaging the public in policy making is one of the most
significant challenges public administrations struggle with. For this reason, although
in PoliVisu pilots this type of relationship is present only in a few cases, it can assume
great relevance in the digital age. In fact, thanks to the development and diffusion of
ICTs, there is an increasing interest in the citizens' broader involvement both as data
providers and as contributors in decision making processes (Calzada 2018). Thus,
public administrations should identify strategies to better structure and develop these
relations which may be crucial for further developing digital innovation.

7.5 Conclusion: Dealing with Complexity in the Era of the Data Shake

Through the analysis of the PoliVisu pilots' direct experience, the article described
some particularly relevant challenges related to the increasing complexity of decision
making ecosystems and the strategies that public administrations can put in place to
take advantage of the data shake. In particular, the causes and effects of the expansion
of decision making networks, the involvement of many different actors with specific
expertise, and the contextual development of new relationships that bind them were
analysed. In this framework, public administrations assume a central role in PoliVisu
for the promotion of digital innovation because, thanks to the activities they are
involved in, they face new challenges and opportunities providing relevant insights.

On the one hand, it is possible to observe how data can lead to a transformation
of decision-makers' activities, effectively supporting every step of the policy cycle,
from the design to the evaluation of a policy. Moreover, data allows conceiving the
policy making process as an experimental activity in which a greater awareness of
current trends described by data can be combined with a renewed ability to manage
and retune the implementation of the measures aimed at achieving a specific policy
goal.

On the other hand, this opportunity strongly depends on the ability of public administrations to rethink their role and tasks in the age of digital innovation, following a *datacy* oriented approach. This process starts with the awareness that the exploitation of data requires an effort to involve multiple actors, both internal and external to the public sector, the structuring of new relationships, and the need to make the internal functioning of the administrative and decision making machine more organic and efficient.

PoliVisu has made it possible to identify these novelties and understand which psychological and structural barriers have been faced by pilots in their path which can keep playing a role in the limited use of data in decision making processes. The analysis of the cases has allowed learning valuable lessons that can help to support the innovative approaches of other administrations outside the project and contribute to the research on digital innovation. For this article's focus, the biggest challenge is the need to build partnerships and relationships both inside and outside the organization and to create the organizational and bureaucratic conditions allowing them to run efficiently.

Considering the internal relations, the pilot cases show that to have an efficient data-supported policy making process, the pilot coordinator (the smart city office) would have benefited from efficient and productive connections with internal actors from all levels (policy makers, operative members) in a regulated framework for the efficient management, sharing and exploitation of available data. Therefore, it is a matter of facilitating all those relations indicated as data and knowledge exchange, internal cooperation for policy making and, if the actors with technical expertise are within the organization, the relations related to data management, analysis, and visualisation. To make these interactions more profitable it is advisable to think in an un-siloed perspective, i.e., breaking the traditional barriers that delimit the different sectors creating cross-cutting collaborative tables starting from the ones that are considered more adapted and readier to a full digital transformation.

At the same time, it is crucial to be aware of the value and usefulness of data in decision making processes, considering them as a precious resource to be wisely collected, managed, and exploited for multiple activities and operations. Both the conditions above must be accompanied by strong political support. An evolution of the governance structure can only be promoted by expressing a strong political will by those who have the power to encourage innovation.

In contrast, external relations introduce a further level of complexity. While internal relations pose an advantage because involved actors know how the organization functions, external relations are based on the participation of new actors who think differently from the public sector. However, PoliVisu has noticed how the construction of private-public partnerships can drive the establishment of efficient collaborations with good advantages for all parties. This opportunity can be seized by overcoming some barriers related to negotiating and defining clear agreements for the collection, use, and management of data. To achieve this outcome, it is paramount to specify which relation to establish with a partner and what to give in return. In line with the evolution of the legal frameworks governing these interactions, the need to build ad hoc procurements to regulate data exchange, access, and ownership, and the extent to which external partners can be involved to offer data-based services to public administrations, will be increasingly pressing. In this way, it will be possible to ensure the contractual balance between the different actors involved in data exchange processes and make interactions between all the actors of the ecosystem increasingly efficient and profitable.

Acknowledgements The author thanks Freya Acar, Joran Van Daele and Jonas Verstraete for their valuable contribution and thorough revision of this chapter, and for their support and availability during the stay in Ghent, and Matteo Satta for the hospitality and the precious inputs during the stay in Issy-les-Moulineaux.

References

Batini C (2018) Datacy, perché una scienza per studiare l'impatto dei dati sulla società in "Agenda Digitale". Available at https://agendadigitale.eu/cittadinanza-digitale/datacy-perche-una-scienza-per-studiare-limpatto-dei-dati-sulla-societa/a. Accessed 5 December 2019

Calzada I (2018) (smart) Citizens from data providers to decision makers? The case study of Barcelona. Sustain 10(9):3252. https://doi.org/10.3390/su10093252

Dente B (2011) Le decisioni di policy. Come si prendono, come si studiano. Il Mulino, Bologna

Kitchin R (2014a) The real-time city? cig data and smart urbanism. GeoJournal 79:1–14

Kitchin R (2014b) The data revolution. big data, open data, data infrastructure & their consequences. Sage, London

Rabari C, Storper M (2015) The digital skin of cities: urban theory and re-search in the age of the sensored and metered city, ubiquitous computing and big data. Camb J RegNs, Econ Soc 8(1):27–42. https://doi.org/10.1093/cjres/rsu02

Thakuriah P, Tilahun NY, Zellner M (2017) Big data and urban informatics: innovations and challenges to urban planning and knowledge discovery. In: P Thakuriah et al. (eds) Seeing cities through big data. Springer Geography

Wiliford C, Henry C (2012) One culture: computationally intensive research in the humanities and social sciences. Council on Library and Information Resources, Washington DC

Giovanni Lanza PhD student in Urban Planning Design and Policy (UPDP) at the Politecnico di Milano where, in 2017, he obtained his MSc in Urban Planning and Policy Design. Besides his doctoral activities, he has been teaching assistant in Master level courses at Politecnico di Milano since 2017. Through his doctoral research, financed by a thematic scholarship (2019–2022) and developed in close cooperation with the Polivisu project partners, he intends to deepen the opportunities related to the use of big data in the evolving frame of mobility studies and transport planning practice.

Chapter 8
Making Policies with Data: The Legacy of the PoliVisu Project

Freya Acar, Lieven Raes, Bart Rosseau, and Matteo Satta

Abstract The PoliVisu project has the goal to investigate the potential of data use and visualisation in urban policy making. The project has explored how data supported policy making is adopted by public administrations and what we can learn from their experience. This is done by enrolling pilot cases with different and specific policy problems. From the experience of the PoliVisu pilots the influence and added value of data in the policy making process is assessed. Considering the recent "shake" in data production and use, PoliVisu has adopted four driving questions, as follow: what are the new roles data can play in the policy making process?, What is the added value of data for policy making? How can innovative visualisations contribute to improve the use of data in policy making processes? To what extent can an increased adoption of data affect the policy making process? How is the data shake affecting the involvement of non-institutional actors in the policy making process? This paper explores these questions, by presenting the experiences and the lessons learnt, also focussing on specific pilots' initiatives and results.

Keywords Data driven policy making · PoliVisu project

F. Acar · B. Rosseau
Dienst Data En Informatie, Bedrijfsvoering, Stad Gent, Belgium
e-mail: Freya.Acar@stad.gent

B. Rosseau
e-mail: bart.rosseau@stad.gent

L. Raes (✉)
Digitaal Vlaanderen, Brussels, Belgium
e-mail: lieven.raes@vlaanderen.be

M. Satta
Issy Média, Issy-Les-Moulineaux, France
e-mail: matteo.satta@ville-issy.fr

8.1   Data Supported Policy Making Through the Eyes of the PoliVisu Pilots

The digital era and the rise of digital data have opened up possibilities for data supported policy making leaving some questions open, including what is data supported policy making, and what does it imply. "Data supported" is a trendy term that is often used and has a different meaning for every user.

"Data supported policy making" hints at a collaboration, almost a symbiosis, between data and policy making. Nevertheless, the way this collaboration works and to what extent this symbiosis is effective, still need to be investigated (see Charalabidis 2021 for an overview of the main barriers for data supported policy making). The PoliVisu project represents a step forward in this direction. "Policy making" is a process focussing on a policy issue, which is often a complex cluster of problems that encompasses many views and has repercussions on a variety of urban dynamics and domains.

Data management should form the backbone of any process that involves collecting, storing and using city data. PoliVisu's experience in pilot Cities and Regions has made it possible to identify the fundamental steps needed to deliver an effective data management strategy. These steps involve taking clear and early actions on data literacy, collection and readiness.

Evidence to support PoliVisu's approach for dealing with data has mainly been collected from five pilots with differing size, data readiness levels and needs, Pilsen (Czech Republic), Issy-les-Moulineaux (France), Ghent, Mechelen and the Flanders Region (Belgium). By exploring different city situations from data, scenario, policy making and administrative points of view, PoliVisu is better able to define a flexible methodology that can be leveraged by a wide range of other cities.

These differences are best illustrated by the example of Issy-les-Moulineaux, which is part of a larger metropolitan area (Île-de-France) and so has to deal with various data sharing and competence issues that are usually not applicable to smaller, stand-alone cities. In the case of Issy-les-Moulineaux, getting the necessary data for its locale requires a very close-knit collaboration with other public bodies and even private companies, e.g. Be-Mobile.

Similarly, for Ghent and Mechelen, the challenge was to obtain the necessary data from third parties as no in-house data was available for the planned use case. This in turn required the city to form partnerships with local service suppliers-cum-data owners. Finally, Pilsen and Flanders Region worked using their own data, but because they are using different sources (Traffic sensors in the first case and ANPR cameras in the second) they have to deal with another issue, i.e. the integration of a large number of data sources.

All the activities of the public authorities have been focused on defining co-creation actions that might make the implementation of data (and related tools) processes more effective.

The experiences of the Polivisu pilots revealed some common challenges that can be synthetized in the following questions:

- what are the new roles data can play in the policy making process? what is the added value of data for policy making?
- how can innovative visualisations contribute to improve the use of data in the policy making process?
- to what extent can an increased adoption of data affect the policy making process?
- how is the data shake affecting the involvement of non-institutional actors in the policy making process?

In the following paragraphs the four questions are discussed in relation to the perspectives adopted by pilots in the experiments and attempts made while searching possible responses.

8.1.1 Data for Dialogue

Using data as a supportive resource for dialogues has different communication aspects and has various impacts on the target groups. PoliVisu tried to manage two challenges. The first was to close the chasm between diverse policy domains in and between government organisations. The second was using data visualisations as a communication tool for policy making to citizens.

The first challenge was to facilitate the communication and the cooperation between diverse policy domains and government organizations. In the pilot of Issy-les-Moulineaux for example the policy issue was traffic congestion. After thoroughly investigating the problem, it became clear that the congestion was not caused by citizens of Issy, rather by drivers working there or transiting to reach Paris. This required the collaboration with various stakeholders, in the public and private sector, which brought to discuss the problem with key actors who need to be aligned on the problem setting, the language and the available opportunities (see Raineri and Molinari 2021). Thanks to data and data visualisations the policy issue can be observed through clear images and dialogue activated towards a shared understanding and solutions development.

However, understanding data and talking about data-related evidences require a certain level of data literacy. By extracting visualisations from the data, the data becomes more comprehensible and improves effectiveness of the communication.

The Flanders accident map managed to set up cooperation on accident data interpretation between the Federal police and the traffic safety institute. This approach had led to a better-streamlined interpretation of the data and cooperation to locate the accident spots better. An online accident map, visualising road accidents, is an excellent tool to give all involved parties in traffic safety insights in where the most serious accidents happen. Also, citizens have the necessary information to elaborate on the local traffic safety situation and discuss traffic safety in their (local) community. The Flanders accident visualisation can be seen as an enabler for local policy making to improve, for example, the traffic situation around schools.

Pilsen used a traffic model as a successful communication instrument for citizens. Traffic models were long considered as an internal tool of the traffic department to perform what-if analysis. An open visualisation interface allowed citizens to elaborate on the impact of the planned road works on their neighbourhood at different moments. Citizens could find out if their neighbourhood were affected positively or negatively. The user-friendly traffic model visualisation was accompanied with practical information about the road work planning, necessity and expected improvements.

Mechelen has an active policy of rolling out schoolstreets. The city of Mechelen wants to know the effect of the schoolstreet on the traffic in the school street itself and on the surrounding streets. Parents were involved to install a traffic counting device (telraam) behind their window and a public dashboard is able to visualise the life situation and to measure the long-term effect of the schoolstreet. The results were used to dialogue with the school community and the surrounding neighbourhood. As a result the expected negative impact as expected by some people living in the neighbourhood wasn't the case.

Issy-les-Moulineaux created, in collaboration with a local startup called MyAnatol, a dashboard to analyse traffic data on the main axes of the city. But the analysis of cars alone was not enough in a period, due to COVID-19 outbreak, that saw an increase in measures in favour of bicycles. This usage has been added to it. The innovative side is also linked to the publication in open data of the data from the dashboard and all the traffic measurements, which are therefore now updated weekly and available to everyone on the data.issy.com portal. This data allowed the City to evaluate its policies about bikes introduced after the lockdown, but also to have a wider view on bikes use, with a comparison between 2019 and 2020.

The four examples above show three different kinds of interaction between stakeholders and citizens: the accident data and map are beneficial for co-creative policy design to improve the local traffic situation; the Pilsen road works communication turned out to be a good communication instrument for policy implementation; the dashboards in Issy-les-Moulineaux and Mechelen turned out to be very useful for evaluation purposes.

8.1.2 Between Precision and Usability

In data supported policy making, data and data visualisations are used to enable discussion between the actors that are involved in the policy making process (Androutsopoulou and Charalabidis 2018). Visualisations, in particular, are essential to facilitate this conversation, especially if supplied through the adoption of tools enabling dynamic visualisations.

Tools allow the gathering of data, the combination of different data sets and different visualisation of data. At different stages of the policy making process different tools can be more or less suitable as each stage may have specific needs (Verstraete et al. 2021) At the early stage of the policy making process, an easy to use tool can be preferred as not requiring special skills or abilities. When more specialized analyses are necessary more specific and powerful tools may be adopted that require data scientists' expertise for the production of more precise, reliable, and effective results. Finally, when the results are discussed with policy makers, again an easy to use tool may be preferred.

In the Flanders and Pilsen road accident maps QGIS was initially used to test the data suitability; later on a BI (Business Intelligence) tool allowed more precise analysis to gain insights from the relationship between accident data and other, more contextual, data sets; finally QGIS was used again for data editing and for the test of the visualisation layout.

The Ile-de-France Region has launched the platform "Ile de France Smart Services". The city of Issy-les-Moulineaux have been between one of the first Cities to have signed this agreement. The implementation of these services is the result of an unprecedented partnership approach around data between public and private actors, as it represents also a common data portal including datasets and data visualisation tools.

Using tools in coherence with the policy making specific needs and sharing their use among different offices of the same organization or with other organizations gives different advantages. First if more local governments adopt the same tool this can significantly reduce the public cost of tool development or acquisition. Second, it also allows for more straightforward communication between public administrations because the tool and visualisations are the same. Also some pre-processed data can be reused since they are already digested by the tool. However, it should always be possible to change and improve the tools in coherence with policy making needs: flexibility and possibility to experiment should be the main characteristics of tool fit for data driven policy making.

> PoliVisu developed the open WebGLayer tool for advanced visualisation tool and the open Traffic Modeller tool for traffic predictions. Both tools allowed Flanders and Pilsen to create tailored visualisations of their traffic safety and accident map and allowed them to open traffic models to the public. The rather limited differences in the data content, made it possible to deploy without much effort tailored visualisations.

Then, from tools and analyses, visualisations arise. Visualisations are an effective way of communication with the variety of people engaged in the policy making process. Visualisations can be used to start a conversation. They can show the situation as it is, where difficulties are situated, and how policy decisions can change and improve the situation. Visualisations are a work in progress. They can continuously be improved to ensure that they are easy to understand and effective in reflecting the policy problem with reduced bias. Constructing a useful visualisation is a challenging and time-consuming task. The use of suitable visualisations responding to specific discussion or dialogues requirements asks for the capacity to balance between precision and usability of the visualisations s as well as for a certain flexibility in the adoption of the best tool.

8.1.3 Proneness to Iterative Process

To move from policy making to data supported policy making asks for data maturity and data literacy in the organization.

As already highlighted by Verstraete et al. (2021) the policy making is structured on to three stages: policy design, policy implementation and policy evaluation. In the policy design stage, the context of the policy problem is assessed, a policy formulation is constructed, different scenarios are hypothesized and analysed and subsequently a policy decision is made. In the following stage the policy decision is implemented, through an implementation plan. The effects of the policy implementation are monitored and apt communication concerning the policy is vital in this stage. In the final stage, when the policy decision has been executed for some time, the policy evaluation takes place through an impact assessment. Based on this assessment a restructuring of the problem might take place, resulting in a new policy problem.

This policy making model is already iterative by nature. After a policy decision has been evaluated it often results in a new policy problem. Furthermore, at any stage in the policy making process it is possible to go back a few steps and restart the process guided by the lessons learnt. This is the key core of the impact of data on policy making: it is transformed into a trial/error process where it is essential to register, reuse and share the outcomes and learnings developed throughout the process (Concilio and Pucci 2021). By virtue of these learnings the data maturity of the organization, the public administration, grows, bringing new insights to the field. These insights in their turn can indicate that it might be opportune to go back to a previous step in the policy cycle. The bigger availability of data augments the option to check the policy measures during the implementation so reducing the risk for irreversible mistakes.

> In Pilsen and Mechelen, thanks to the availability of a traffic model and a list of the planned road works, the impact of multiple planned road works and deviations can be simulated. In Mechelen, a service was tested to simulate the immediate effect of the occupation of the public domain as part of the approval procedure, including signalling plan and electronic payments.

Furthermore, depending on the step and stage of the policy cycle different issues arise (Verstraete et al. 2021 for a complete overview of the types of data analysis in each step of the process). In the first stage, the policy design stage, the policy and data issues are more related to the scope of the policy problem. By setting up a context and through the analysis of scenarios data has a guiding role in this stage, allowing more insight in the context of the policy problem. In the policy implementation stage the questions are more related to monitoring and fine-tuning. In this stage data is used to follow up on the policy decision and adapt or adjust the policy decision if necessary. In the final stage data use is more related to analyses and checks, whether the policy decision was successful, sufficient, and how it affected the domain of the policy problem and other, related domains.

Additionally, data has the property of moving quicker through the policy cycle than the policy itself. In consequence the results and learnings from the data can have an influence on the policy making process at any given point and require flexibility and an agile performance from the actors involved in the policy making process.

8.1.4 Actors Involved in Data Supported Policy Making

Data supported policy making is a complex process that involves many different actors. A network analysis of the PoliVisu pilots showed that at least 20 actors were involved in every pilot (see Lanza 2021). This includes partners both internal and external to the organization. For every pilot a core group could be distinguished from peripheral partners. It is clear that the core group is a multidisciplinary team where people with varying expertise work closely together.

Agile and effective data-supported policy making is an interdisciplinary challenge and requires the combination of multiple perspectives (Walravens et al. 2021). At least a policy perspective and a data perspective have been identified. From the data perspective different competences are required. These different competences are hardly found in one multi-talented person. Moreover, one person will never have the time to deal with the multitude of tasks related to data-supported policy making. On the other hand, most organisations will not be able to install a complete multidisciplinary team at once. A good starting point is to focus on three general profiles: a data-engineer to cover data management and development needs a general data-analyst to cover the data-analyst and data science needs; and a policy (decision) maker. Gradually, with the evolution to more complex analyses and a more mature data-supported policy making, the organisation can invest in specialized roles such as expert data-analysts, statistician, data scientists and researchers.

In the Ghent pilot several (internal and external) partners are at work to handle the policy problem. A data-engineer is at work to maintain the datasets and provide a framework to work with the datasets. Data-analysts are working within the public administration at the office of data and information and externally at the telecom provider. Finally, a close collaboration exists between the office for data and information and the policy makers involved in the student housing problem setting.

In Issy-les-Moulineaux, the City created a dashboard and related KPIs to connect the various departments with policy makers and, at the same time, providing a simplified version for citizens to use data to help them to have better information about the impact of policies.

A data-driven organisation must ensure through its organisational structure, the collaboration of the different actors in the policy making process. Related to the data activities, three main organisational forms can be distinguished: a centralized organisational structure, a decentralized structure and a balanced hybrid form between these two.

In a decentralized organisation, business or policy domain units develop their own data analytics teams. This promotes the responsiveness of the dedicated data teams to the priorities of the units. Also, since the data teams are embedded in a policy domain, they are likely to develop thorough domain knowledge. However, isolated decentralized teams might suffer from siloed data expertise, the inability to develop an organisational data strategy and the lack of broader managerial focus. Smaller decentralized teams probably will not have dedicated data engineers and developers. For ad hoc analyses this might not be a great concern, but the analysts will be unable to deploy relevant results to productional data-pipelines and automated analyses. Data analysts might also struggle with flexible data access and the deployment of generic analysis and visualisation tools.

A centralized structure has many advantages in terms of talent and knowledge management, the potential to develop a cross-departmental data strategy and a broader managerial focus. Still, a central unit might face important challenges concerning the allocation of sufficient resources to individual business units and flexible responses to domain priorities. Centralized data teams need to invest extra time to gather sufficient domain knowledge. The installation of a centralized data team can be a good starting point to engage in data-supported policy making. Sooner or later, organisations will feel the need to evolve to a more hybrid form to balance the advantages and challenges of both the centralized and decentralized organisation structure.

Because of the specific knowledge required by working with (big) data for policy making it happens that the data experts working for the data providers become effective collaborators of the policy making organization so transforming it into an hybrid structure.

The city of Ghent started working with the data scientists from Proximus to ensure a reliable data analysis. This collaboration showed to be a win-win process for both parties. The city learned how to work with this kind of big data sources and Proximus learned how a city operates and how it formulates its requirements to (big) data.

Issy-les-Moulineaux collaboration with a local startup, My Anatol, and its data specialists. This made possible for the City to access data and skills and the startup to improve its offer for public authorities, being able to have a real proof of concept in real, through the various requirements and feedback received from the City.

Actors involved in the policy making process are different: policy makers, operative sectors of the public administration, an office for data and information within the public administration, technical service providers, data providers and the public. Every actor has a specific task or purpose. It is highly relevant to identify which policy makers you need support from, which operative sectors of the public administration you require information and who in the office for data and information that can aid with the data management plan and find the link between data and the policy issue. External to the organization, other actors might be necessary either because they are data owners or because they can help with data analysis and visualisation. The pilots realized that identifying and interacting with diversified actors since an early stage of the policy making process augment the productivity of the process.

In conclusion, for data driven policy making, collaboration between many different actors, internal and external to the organization, is required. Because of the complexity of the process, it is advisable to set up structures and come to agreements at the start of the process.

8.2   Bottlenecks and New Practices Detected in Policy Making

The project showed that the way to the full exploitation of data potentials in policy making is still long. PoliVisu pilots experienced several bottlenecks in data availability and use that still represent important limitations to effective processes of data driven policy making. Such bottlenecks, managed by the project pilots, highlighted important adaptation that gave rise to new use of data.

The findings mentioned in this section have been collected through the observation on various pilots mentioned in the first section of this paper. This work made the actual development, implementation and monitoring of local policies in four cycles, feeding the results back into the overall project's solutions.

8.2.1   Bottlenecks

The different pilots of PoliVisu started the project with a high degree of knowledge about open data and work with simple datasets. At the same time, they had ambition, but a limited knowledge on use of big data and smart visualisations.

Those pilots entered then into the project with some interesting scenarios, but those were created without a real certainty about the possibility to really deploy them successfully.

One of the questions that PoliVisu wanted to answer with the support of practical cases, at least for local public authorities, is "Do we really have a resistance to innovation processes in the public local authorities?". According Ritchie (2014) the theories claiming an anti-innovation approach of governments have somehow a base of truth, but he also considers that it is still needed to have a deeper research to understand the reasons.

Table 8.1 mentioned bottlenecks, detected all along PoliVisu, seem to show how the resistance is somehow structural and often not just related to individuals.

Table 8.1 Bottlenecks faced by the PoliVisu project

Data literacy	According to the Data Literacy project, "76% of key business decision-makers aren't confident in their ability to read, work with, analyse and argue with data". The PoliVisu project investigated the issue of data literacy using its own survey which made confirmed this issue. At the same time, this was underlined by most of the pilots (all, but Ghent), at the beginning of the project, which declared not to have data scientists or analysts in their teams
Data ownership	All pilots had to deal with data ownership, but in two cases this is particularly true. Ghent and Issy-les-Moulineaux found themselves in a situation in which they had the obligation to get the data from private bodies, consequently, to have a real work to benchmark the market and to find affordable solutions. This is extremely time consuming and it plays a, potentially negative, role in deployment of data projects
Data fragmentation	Often the various municipalities and/or other public bodies do not share data or do not have compatible formats and policies. This is a real point as it makes data often non exploitable and it was experiences by all the pilots of the project
Fragmentation of jurisdiction	The fragmented data is also related to the division of competences between various bodies. This was particularly impacting in Flanders and Issy-les-Moulineaux due to their geographical configuration as Flanders has to deal with various Cities and Issy-les-Moulineaux is part of a big urban agglomeration which is very dense, and it has many decision levels
Data privacy	Data privacy policies oblige, particularly with the new GDPR regulation, to be extremely careful to any potential breach of data privacy. In particular, the case of Ghent in which mobile data was used is key. Even if the data was anonymized, the detection of a person can represent a real issue, consequently important steps need to be taken with, next to anonymization, aggregation of data

The bottlenecks listed above are all obviously important, but there are some having somehow a heavier impact.

In particular, the PoliVisu experience in its Cities and public authorities shows that private companies often hold the most useful datasets, as often public data (and competences), is fragmented. This evidence shows an unbalanced relation between the private and public sector which may explain, at least partially, a reluctance to innovate in data in many cities and public authorities.

8.2.2   New Practices and Knowledge

As it is now clear to everyone, an increasing reduction of specific budgets of Cities
(including the shift of priorities), and public authorities in general, is today a reality.
This reduction cannot be tracked with a clear trend, being highly different country by
country, but it follows the trend of national public budgets. The Council of Europe,
already in 2011, was warning about potential negative effects, even recognising the
need of contribution of public authorities, in an ad hoc publication.

As explained in the previous paragraph, the PoliVisu experience and the ones of
its Cities and Regions show how a data transition requests high level investments, not
without risks, for the public authorities both from a data and visualisations point of
view. One of the major findings of PoliVisu, as reported above, has been the detection
of an unbalanced relation between the private and public bodies, being often the most
valuable data held by private companies.

In this framework, PoliVisu pilots showed how it is possible to move through this
bottleneck and the ones mentioned above, partially or totally, adopting some new
practices, summarized in Table 8.2.

Those practices, collected through observation and feedback received, made
possible to highlight how the process in adoption of data in policy making has an

Table 8.2 Emerged practices

Awareness of data value	The various pilots were already aware of the value of data in itself, having already advanced open data strategies in place and a vision related to data. At the same time, in particular at decision makers level, this knowledge was just guessed. During the project, the meetings with policy makers have totally unlocked the potential of pilots that could start working 100% on such projects with a snowball effect. The various pilots that came out or could be deployed anyway in Issy, Mechelen and Pilsen in COVID-19 time are really an example of how pilots took data as a real resource
Use of dashboards	The pilots at the beginning of the project had quite interesting ideas about tools and solutions, but those were theoretical and not often realistic. As the project moved on, pilots had to deal with reality, and they could finally find solutions. Every pilot has its own specifications, but a common point was the use of dashboards to have analyses of data. Those tools are useful to connect the policy makers with their operational departments and, in some cases, directly to citizens to make better accepted and/or understood some policies
Co-creation projects	The PoliVisu's cities have started the project with scenarios that looked "self-standing", but the various bottlenecks met and PoliVisu itself pushed them to involve more and more other stakeholders. Finally, all pilots had constructed co-creation projects in which other private and public partners played a role. This was extremely evident in Ghent, Issy and Flanders, but absolutely true also in Pilsen where applications created by local start-ups, universities and associations were mixed with the City open data portal to improve the information

upward trend as Cities and public authorities may have a slow start, but, as soon as they start, they will be improving quicker and quicker with time. The improvement will also ameliorate the communication to citizens, making also possible to unlock more and more funds, which will support the whole process.

To do so, it is really important to engage the City in co-creation processes and projects which make easier to unlock this potential, giving access to the City to data and skills that would, otherwise, never be available.

8.3 Conclusions

Through its pilots and their stories, PoliVisu has showed the potentials of Open and Big Data in policy making. This closing section aims at wrapping up the whole stories of the project, giving also some recommendations to deploy, in cities, projects through the use of data.

8.3.1 Lessons Learnt from the PoliVisu Project

In data supported policy making, data is used to commence a dialogue. Data can activate dialogues, public dialogues, concerning a policy problem, and can support arguments and visions concerning the challenge at hand. Adding data and data visualisation to the conversation allows for a better understanding of the problem, the context, and the possible effect of policy decisions. However, data can support dialogues about policy issues without neglecting the complexity of the policy process where many viewpoints and approaches are intertwined. Vision, knowledge and experience still represent the fundament of policy making, and data can contribute to a part of that knowledge.

Getting knowledge from data is a challenging process. Visualisations aid everyone involved in the dialogue to understand the data. We distinguish between three types of visualisations based on the stage of the policy making process in which they are used. At the beginning, when only an exploratory analysis is performed and the dialogue concerns preliminary results, visualisations are used that are accessible and easy to use. The goal is to transfer, explore and discuss intermediate analysis results. This allows for the exploration of some general trends, but no in-depth analysis and results can be displayed.

Then, when the results are analysed more in depth, flexible and powerful visualisation tools may be more appropriate. These visualisations require expertise from data scientists. The visualisations allow the data scientist to better understand the data and obtain results that would not be visible with more general visualisations.

Once the in-depth analyses have been performed, visualisations are developed that allow policy makers and a broader audience to query the data. Easy to use visualisations support policy makers when monitoring.

All these visualisations types should be flexible and evaluated at certain points in time. New issues might arise, and new information might come to light. It is important to note that it is not necessary to stick with a visualisations type once one has been chosen. It is always possible to change the visualisations that need to be flexible to best support the policy making process.

The Polivisu project also showed how data supported policy making is an iterative process. At every stage of the policy making process it is possible to go back and start again by considering the lessons learnt. Likewise, the data activities related to data supported policy making happen in an iterative manner as well. As shown in the policy-oriented data activities framework (as described in Verstraete et al. 2021) there is a continuous process where analysis results go back and forth between the policy maker and the data scientist with the ultimate goal to obtain sensible results that can be used to support policy making through close collaboration between the policy maker and the data scientist.

Data supported policy making is a team effort that requires specific and varying expertise from the people involved. In short, data supported policy making is a complex and challenging process that requires communication and collaboration between a diversified group of actors. These actors, and the organization, will gain data maturity through time. This might come across as a slow process, but every small change has a significant value.

8.3.2 Some Recommendations

The PoliVisu project highlighted some pathways for the situation to be improved towards a more effective integration of data in policy making. They are all illustrated in the following final Table 8.3.

Table 8.3 Recommendations from the PoliVisu project

Increasing Data Literacy	As reported above, Data Literacy is a real blocking point to deploy a good strategy. To this end, it is necessary to hire and to engage data analysts and scientists, at least as subcontractors. This is a key point to be successful
Breaking Silos	Large and medium sized municipalities, in fact, are normally "siloed" structures, often not well connected between them. In this sense, even the data produced or stored by these silos are considered as a sort of exclusive property, which is not shared with other silos, even if that may bring benefit to the Municipality as a whole. Siloed organizational status is reproduced in data production and management. The best way to tackle this barrier is to create cross-cutting working tables with various services, starting from the ones that are considered more adapted and ready for a full digital transformation
Showing data value to key internal players	The second relevant element of the political culture affecting the management of data is related to the role played by data in the Municipality procedures. Data is rarely, almost never, considered as a useful resource per se; it is rather seen as a functional component of bureaucratic procedures and, as such, not considered as a relevant output of any process. This reduces the attention to data production and management and does not include any scenario of data re-use or utilization in other activities or processes. It is clear that failure in considering data as public good finds its origin in the (merely) bureaucratic approach to public service production and supply; one could even say that public services themselves are not considered or managed as common goods. To move through this barrier, it is needed to show the value of data, creating some first useful applications in a pilot mode, the positive reactions of citizens and external stakeholders will represent a real motivation for City teams

(continued)

Acknowledgements Special thanks to Jonas Verstraete and Joran Van Daele for their valuable contributions and thorough revision of the present document.

Table 8.3 (continued)

Increasing Data Literacy	As reported above, Data Literacy is a real blocking point to deploy a good strategy. To this end, it is necessary to hire and to engage data analysts and scientists, at least as subcontractors. This is a key point to be successful
Giving a strong political support to digital transformation teams	The third element is strictly related to individual behaviours, being a project to implement data a real change of paradigm that requests a strong effort in the short term. It is not obvious to have teams of the various departments to "hide" their non-effort to make the internal procedures improved through data. This psychological effect is actually related to the same reasons related to the behaviour shift in mobility, while a person that for 20 years has used the car to go to work, even when not motivated enough, refuses to change, even if confronted with clear proofs that a switch would give him/her an advantage in the medium-long term. The PoliVisu's Cities experiences showed how a strong political will is absolutely necessary to go through this resistance
Building partnerships with the private sector	PoliVisu has noticed how the construction of private-public partnerships, also with some minor financial contributions of Cities and other public authorities, can drive to the construction of efficient collaborations with good advantages for all parties. Actually, the project could stimulate Cities to look for private partners providing data (Issy) or tools (Pilsen) or both of them (Ghent) and to settle an on the ground collaboration. While this collaboration starts, the project has noticed how Cities start a quick innovation process, showing how the usual anti-innovative approach can change, and private companies show an unusual capacity to support them, also providing, in some cases, investment (Issy-les-Moulineaux and Ghent particularly)

(continued)

References

Androutsopoulou A, Charalabidis Y (2018) A framework for evidence based policy making combining big data, dynamic modelling and machine intelligence. In: Kankanhalli A, Ojo A, Soares D (eds) Proceedings of the 11th international conference on theory and practice of electronic governance, Galway, Ireland, 4–6 April, pp 575–583

Charalabidis Y (2021) Policy-related decision making in a smart city context: the PoliVisu approach. In: Concilio G, Pucci P, Raes L, Mareels G (eds) The data shake. opportunities and obstacles for urban policy making. Springer, PolimiSpringerBrief

Table 8.3 (continued)

Increasing Data Literacy	As reported above, Data Literacy is a real blocking point to deploy a good strategy. To this end, it is necessary to hire and to engage data analysts and scientists, at least as subcontractors. This is a key point to be successful
Including data clauses on public procurements	Cities need to show a good capacity to learn from their past mistakes. In particular, the lack of inclusion of clauses in public procurement contracts is one of the biggest lessons learnt from Cities, making those clauses, from now on, fundamental in all public tenders. It is then absolutely necessary to include clauses on public procurements to access the data and, when necessary, to have included also a good format making it quickly usable. Cities should also consider whether, in addition to getting access to the data themselves, the contracts should require the supplier to make the data available as open data or to other private sector actors on a fair and equitable basis so that innovation and societal benefit can be maximised

Concilio G, Pucci P (2021) The data shake. An opportunity for experiment-driven policy making. In: Concilio G, Pucci P, Raes L, Mareels G (eds) The data shake. opportunities and obstacles for urban policy making. Springer, PolimiSpringerBrief

Lanza G (2021) Data related ecosystems in policy making. The PoliVisu context. In: Concilio G, Pucci P, Raes L, Mareels G (eds) The data shake. opportunities and obstacles for urban policy making. Springer, PolimiSpringerBrief

Raineri P, Molinari F (2021) Innovation in data visualisation for public policy making. In: Concilio G, Pucci P, Raes L, Mareels G (eds) The data shake. Opportunities and obstacles for urban policy making. Springer, PolimiSpringerBrief

Ritchie F (2014) Resistance to change in government: risk, inertia and incentives. University of the West England, Economics Working Paper Series 1412

Verstraete J, Acar F, Concilio G, Pucci P (2021) Turning data into actionable policy insights. In: Concilio G, Pucci P, Raes L, Mareels G (eds) The data shake. opportunities and obstacles for urban policy making. Springer, PolimiSpringerBrief

Walravens N, Ballon P, Van Compernolle M, Borghys K (2021) Data ownership and open data: the potential for data-driven policy making. In: Concilio G, Pucci P, Raes L, Mareels G (eds) The data shake. opportunities and obstacles for urban policy making. Springer, PolimiSpringerBrief

Freya Acar Project manager for European and Flemish projects concerning (open) data, data driven policy making and smart city for the city of Ghent (Belgium). She obtained her MSc in Theoretical and Experimental Psychology from the University of Ghent in 2014. Hereafter she started a PhD at the Department of Data-Analysis concerning the assessment and correction of bias in neuroimaging studies. Communicating data questions and results to peers with less data affinity was one of the key aspects of her PhD, which is further explored through visualizations in the PoliVisu project.

Lieven Raes holds master degrees in Administrative Management and land-use planning. Lieven is a public servant at Information Flanders (Flemish government) and is currently the coordinator of two EU H2020 projects regarding the relationship between data, policy making in a smart city context (PoliVisu and Duet). Before Lieven participated in several other EU projects (FP7, FP5 and FP4) and also in several Flemish ICT and E-Government projects as the first Mobility plan for Flanders, and the digitisation of the building grant.

Bart Rosseau has a background in political science and 25 years experience in the civil service. He started the open data programme in the city of Ghent, and is currently heading the Data and Information Unit of the Ghent City Council. He was chair of the Data working group of Eurocities and the Knowledge Society Forum of Eurocities. He is cofounder of Open Knowledge Belgium, and boardmember of the Council of Cities of the Open and Agile Smart Cities network (OASC). Over the years he participated in international projects focused on data in and beyond the Smart City context (SmartIP, OASIS, SCORE, MUV, …), and participated as speaker and panelist in national and international conferences on Open Data, Datapolicies and the impact of data on policy making.

Matteo Satta is a Project Manager, mainly interested in digital innovation and European Union (H2020). He graduated in High School in the U.S. and in International Political Sciences in Turin (Italy). Since 2005, he has contributed to the management and the development of various ICT International projects, such as e-Photon/One and the Researchers' Night in Turin (Italy), and IPR Licensing programs, such as MPEG Audio (MP3) and DVB-T. In 2014, he joined Issy-les-Moulineaux to manage and develop the City participation at EU and international level, with a particular interest on Digital Innovation in Smart Cities. He has specialization in Smart mobility solutions and (open) data.

Acknowledgments

In late 2016, half a year after my on open transport data project ended, colleagues and I realized that building an interactive open data portal is not sufficient by itself to drive the use of data for policy making. Based on this realization, together with Susie Ruston McAleer, Andrew Stott, Hugo Kerschot, Jiri Bouchal, Anna Triantafillou, Bart Rousseau, Dirk Frigne, and Geert Mareels the concept of PoliVisu was born. Beyond making data available, PoliVisu chose to focus on data literacy through easy-to-read visualizations using open and big data, thereby ensuring policy making is understandable by all.

This focus on open and big data for policy making through smart data visualizations would enable more citizen involvement in policy and increase data literacy across Europe. Thanks to an EU Horizon 2020 co-creation call, we were able to realize our views on the terrain. Fifteen partners from Belgium, Czech Republic, France, Greece, Italy, and the UK cooperated on an innovative EU proposal submitted in February 2017. Later that year we received the positive news about our selection and in November we had our kick-off meeting in Ghent, together with Pilsen and Issy-Les-Moulineaux two of the Pilot cities.

While I am writing this preface, PoliVisu is in its final stage. After three years of intense cooperation, we have implemented our approach in six different locations, created a practical toolbox, and participated in several events despite the corona outbreak. Thanks to the flexibility of the consortium, we managed to write this book, the Data Shake, created an interactive Massive Open Online Course (MooC) and designed innovative re-usable open software tools for data visualization. These results were only possible with the support of a great team of project partners. It has been a pleasure to work with them all in a cooperative and ever-innovative way, and I would like to make these thanks public.

Thanks to Gert Vervaet, Geert Mareels, Jurgen Silence, and Bart Scheenaerts from Information Flanders who have helped to coordinate the PoliVisu consortium and drive the technical work. Thanks to Jiri Bouchal and Hugo Kerschot from IS-Practice who took care of all the meetings and administration. Special thanks to Gert and Jiri,

G. Concilio et al. (eds.), *The Data Shake*,
PoliMI SpringerBriefs,
https://doi.org/10.1007/978-3-030-63693-7

for assisting the consortium from the beginning and taking care of the deadlines and quality control.

Thanks to Prof. Grazia Concillio, Prof. Paola Pucci, and Giovanni Lanza for the scientific support throughout the project. Special thanks to Grazia for all the effort to coordinate this book and bringing together the authors.

To Susie Ruston McAleer, Pavel Kogut, and Laura Gravilut from 21C, thank you for all the support from writing the proposal to delivering communications throughout the PoliVisu project and developing the MooC. No effort was too much for you in promoting the project via different channels.

Without the open and big data specialists from our partners HSRS, SenX, Geosparc and Innoconnect, Karel Charvat, Tomáš Řezník, Fabien Tence, Jeroen Saegeman, Pepijn Viane, Kris De Pril, and Jan Jezek it wouldn't have been possible to test our concepts and ideas on the ground in real life.

Thanks to our mobility data experts from P4All, EDIP, and MACQ, Karel Jedlička, Daniel Beran, Jan Martolos, Geert Vanstraelen, and Rob Versmissen who brought together their experience in mobility data, transport modeling and smart camera data; to our Greek partner ATC who shifted from a technical integrator role toward an interactive data dashboard developer and web developer. Thanks to Marina Klitsi, Padelis Theodosiou, and Stamatis Rapanakis for their flexibility and cooperative spirit.

The PoliVisu Pilots in Ghent, Pilsen, Issy-Les-Moulineaux, Flanders, Mechelen, Voorkempen were essential to make PoliVisu an interactive co-creative and inno- vative project. Without the input of Bart Rosseau, Joran Van Daele, Freya Accar and Jonas Verstraete from Ghent, Stanislav Stangl, Václav Kučera and Tomáš Řehák from the city of Pilsen, and Eric Legale and Matteo Satta from Issy-Les-Moulineaux, it wouldn't have been possible to test the PoliVisu concepts with real citizens, city managers, and politicians. Thank you for promoting the PoliVisu project in your city, other partner cities and international fora. Also thanks to Dimitri Van Baelen, Veerle De Meyer from the city of Mechelen and Geert Smet from the Policezone Voorkempen for their support and cooperation.

During the PoliVisu project, the outcomes were also reviewed by our critical friends Andrew Stott (Former UK open Data Manager), Yannis Charalabidis (Prof. at the University of the Aegean), Bart De Lathouwer (CEO Open Geospatial Consor- tium), Nils Walraevens (IMEC—Vrije Universiteit Brussel), and Eddy Van Der Stock (CEO Linked Organisation of Local Authority ICT Societies). Andrew, Yannis, Bart, Nils, and Eddy, thanks for your critical support during the project. Your insights and comments were very helpful to all partners in the consortium. A special thank you goes to Giorgio Costantino, our EU Commission project officer for his helpful advice and strategic guidance.

This book not only reflects the lessons learned from the consortium itself but also the expert views of different field experts. I'd like to thank all the experts who contributed to the Data Shake, Prof. Pieter Ballon, Mr. Nils Walraevens, Koen Borghys, and Mathias Van Compernolle from IMEC, Mr. Petter Falk (Karl- stad University), Paolo Raineri, Francesco Molinari, and Prof. Yannis Charalabidis (University of the Aegean).

Last but not least, I want to thank my wife Inge and son Milan for their love and support during the entire PoliVisu project. It was a pleasure to be surrounded by everyone mentioned throughout this challenging initiative, and I hope the reader will enjoy this book and appreciate the outcomes of the PoliVisu project in the same way I enjoyed working on it.

Lieven Raes

Coordinator PoliVisu project, AIV—Digitaal Vlaanderen

Printed in the United States
By Bookmasters